SOCIOLOGY AND INDUSTRIAL LIFE

Other Michael Joseph books in the TUTOR BOOKS *series*
Series Editor: RONALD FLETCHER

The Making of Sociology, Volumes 1, 2 *and* 3
by Ronald Fletcher

Karl Marx: Economy, Class and Social Revolution
edited and with an introductory essay by Z. A. Jordan

Herbert Spencer: Structure, Function and Evolution
edited and with an introductory essay by Stanislav Andreski

Max Weber: The Interpretation of Social Reality
edited and with an introductory essay by J. E. T. Eldridge

John Stuart Mill: A Logical Critique of Sociology
edited and with an introductory essay by Ronald Fletcher

Deviance and Society
by Laurie Taylor

Also by J. E. T. Eldridge: *Industrial Disputes:
Essays in the Sociology of Industrial Relations.*
Routledge and Kegan Paul 1968

SOCIOLOGY AND
INDUSTRIAL LIFE

J. E. T. ELDRIDGE

MICHAEL JOSEPH LONDON

First published in Great Britain by
MICHAEL JOSEPH LTD.
*52 Bedford Square,
London, W.C.1*
1971

© 1971 by *J. E. T. Eldridge*

All Rights Reserved. No part of this publication may be recorded, stored in a retrieval system, or transmitted, in any form or by any means, electronic, mechanical, photocopying, recording or otherwise, without the prior permission of the Copyright owner

SBN 7181 0740 2

*Set and printed in Great Britain by
Unwin Brothers Ltd, at the Gresham Press, Woking,
in Imprint type, eleven point leaded, and bound by
James Burn at Esher, Surrey*

ACKNOWLEDGEMENTS

My thanks are due to the following for permission to quote from works in which they hold the copyright: Free Press, E. Durkheim, *The Division of Labour in Society*; Lawrence and Wishart, K. Marx, *Economic and Philosophical Manuscripts of 1844*; Methuen, J.-P. Sartre, *Being and Nothingness*; and Bedminster Press, M. Weber, *Economy and Society*.

On a more personal note I am glad to acknowledge my thanks to the following: Ronald Fletcher, the editor of this series, for his extremely helpful advice; Carol Blunt who has dealt with this manuscript for Michael Joseph's with great efficiency; and to Janet Wright for typing the manuscript with much skill and patience.

Since a great deal of this book discusses issues relating to the division of labour, I am perhaps particularly conscious of the many people not known to me personally on the publishing and printing side of the work who have made the production of this book possible. To these also I express my thanks.

CONTENTS

Acknowledgements v

Introduction by Ronald Fletcher 5

Abbreviations

Preface 7

PART ONE *Industry and Society: Theoretical Perspectives*
  (1) Industry and Society: Some Developmental Themes 13
  (2) Systems Analysis and Industrial Behaviour 25
  (3) The Social Action Perspective 40
  (4) Social Action and Reference Group Analysis 64

PART TWO *Industrial Society: Anomie—and a New Integration?*
  (1) Durkheim: Anomie and Economic Life 73
  (2) Hobhouse and Liberal Socialism: A Durkheimian Parallel 92
  (3) The Breakdown of Community as a Form of Anomie 100
  (4) Social Order, Anomie and Industrial Relations 112
  (5) The Merton Anomie Paradigm: Applications and Critiques 119

PART THREE *Individual and Society: Alienation—and a New Humanity?*
  Introduction
  (1) Work and Alienation in Marx 139
  (2) Alienation, Occupational Role and 'Bad Faith' 158

(3) Max Weber and the Issue of Alienation in Industrial Society — 165
(4) Trade Unions and Bureaucratic Control — 175
(5) Alienation and Freedom: A Critique of the Blauner Thesis — 183

PART FOUR *Conclusion*

Conclusion — 197

Bibliography — 209

Index — 222

Introduction by Ronald Fletcher

Professor Eldridge's new book is one of the first volumes to mark a distinctive development in *The Making of Sociology Series*. So far, this has concentrated upon the contributions of leading theorists to the making of sociological analysis as such. Now, these new volumes clarify the application of this analysis in special fields of study. They are, in short, basic texts on sociological *specialisms*—but with an extremely important emphasis. Specialisms are decidedly *not* seen as 'special sciences' examining 'separate parts' of society. The underlying persuasion—deeply shared by editor and contributors—is that there is only *one* sociology, not *many*, and that *specialisms* within it are the application of sociological analysis as such to theoretically distinguishable areas of investigation and enquiry, but areas which are, nonetheless, only properly to be seen and understood within the context of society as a whole. They are all parts of sociology as an entire discipline. They are all distinctive parts of the making and enriching of the science of society as a whole.

The earliest volumes have been selected because of their direct relevance to immediate and pressing problems, so that the books— besides providing texts for use in specialist courses—may demonstrate the truth and value of the sociological perspective by illuminating the many dimensions of these particular areas of social life. No problems lie more crucially at the heart of the changing turmoil of contemporary society than those of conflict, power, and control in the organization of industry—the field of industrial relations— and this book clarifies the contributions of sociology to the study, discussion, and possible solution of these problems. It is an introductory text in *Industrial Sociology*.

Sometimes, however, 'Industrial Sociology' is caricatured as 'Plant Sociology'—concerned only with relationships on the 'Shop Floor'. It tends to be thought of in terms of the Hawthorne Experiments, and their (never-ending) subsequent qualifications.

A*

Even a little wider than that, some industrialists and politicians, as well as some social scientists—especially economists—sometimes speak as though *the economy*, or *the economic system*, was a part of society clearly distinct from all its other social institutions. The organization of production; the distribution of wealth and property; problems of management, industrial arbitration, industrial relations; all these are thought to constitute a separate area of interests and organization—and Economics and Industrial Sociology are the appropriate specialisms for the study of it. Wider matters—of the activities of families, of public welfare and public services, of political power and policies, of religion, of morality in society—are all elements 'extra' to the economic sphere, and obtrude upon it. Indeed, some think that considerations of them should not be allowed to 'distort' the free adjustments of the 'market-situation'.

Sociology, however, has never recognized such separatism. Theorists of the most varied approaches—Marx, Le Play, Weber, Durkheim (to mention only a few), in their studies of the forms assumed by families and communities, of property relations and conflicts of power, of the part played by the religious ethic of a society, of the division of labour—have always seen the practical activities of men (the regulation and organization of work, of property, and economic relationships) as being at the heart of society as a whole: interpenetrated by, and interpenetrating, every aspect of it. Industrial relations have always been seen as being vitally connected with the nature of relationships in the family, law, government, morality—indeed, in the whole order of society. Industrial sociology is therefore the application of this much wider sociological perspective: analysing not only the pattern of industrial activity itself, but also the wider pattern of life—of values, ideas, beliefs, as well as social, legal, and political institutions—which is intimately associated with it.

This point deserves very considerable emphasis. It carries more truth than the simple statement of it suggests. As soon as one considers the work of all the major theorists, it is at once clear that their *central* concern was the disruption of human relationships by industrial capitalism, and a desire to work out some satisfactory kind of social reconstruction. It was not only Marx who saw economic activity and property relations at the heart of the crucial conflicts of power in society. It was not only Marx who despised 'bourgeois' values. Tönnies saw the calculated, rational, contractual nature of bourgeois society as being destructive of the many-sided personal nature of man in natural community. Durkheim saw the whole moral and psychological malaise of modern industrial society in terms of the changed nature of the division of labour, and the ungoverned extension of unsatisfactorily based contracts (i.e. contracts entered into on a basis of inequality.) Le Play saw modern industrial and urban life as being disruptive of community, and of any stability and dignity of human relationships within it. And it was not only one or two theorists like Hobhouse who wrote of this in terms of morality and justice. *All* these theorists did! And some of those engaged in present-day controversies still find it hard to believe that a satisfactory order of 'industrial relations' can be attained until an entire social order is achieved which, in general, is ethically justifiable. The remaking of industrial relationships is seen at the heart of the remaking of society: so that we are back, again, with our concern for, and vision of, society as a whole; with our perspective of sociological analysis as a whole.

Professor Eldridge—himself an industrial sociologist—has provided a book which successfully introduces a very wide range of recent and present-day empirical investigations, and which yet retains and makes clear all these larger dimensions. Marx, Durkheim, Weber, Hobhouse, mingle with Merton, Wright Mills, Blauner, and the Hawthorne experiments!—in these pages, and all these enquiries are drawn together by showing how they circle about perennial and focal concerns—such as the basic notions of 'anomie' and 'alienation'. Problems of theory and research are discussed—but always in relation to specific investigations and crucial human problems. An extensive bibliography is also added for detailed further reading, and for selective use in the working out of specialist courses. The book is therefore an introduction and guide, condensing and relating much information, and throwing useful perspectives over the entire field of study.

Abbreviations

AJS   American Journal of Sociology.

ASR   American Sociological Review.

BJIR   British Journal of Industrial Relations.

BJS   British Journal of Sociology.

IS   International Socialism.

J.Soc.Psy.   Journal of Social Psychology.

Preface

Sociology is an untidy science. The world is its parish and in an empirical sense its practitioners are confronted with endless and continuing possibilities. In consequence their work sprawls out in many directions. One 'solution' to this form of untidiness is to carve up the work into a number of sub-disciplines: the sociology of religion, education, law, politics, industry are common examples. The solution has a certain bureaucratic appeal when it comes to teaching the subject. It also reflects the fact that a good deal of detailed research has been undertaken in these areas with which one has to come to terms. At the same time there is a risk that the sub-area will be treated as a self-contained entity. When that happens, only in a very attenuated way is the original concern of sociology to indicate the inter-connectedness of the whole of social life retained. From time to time this criticism has been made about industrial sociology. It has, for example, been pejoratively described as 'plant sociology' with the implication that the research focus is upon work groups and factory structures with no connecting links to the wider social reality. In *Sociology and Industrial Life* part of my purpose has been to show that industrial sociology need not, and for that matter should not, conform to this image.

Sociology is also an untidy science in the sense that it is littered with concepts, classificatory systems, typologies, models and theoretical orientations. This is reflected in sociological studies of industrial life as elsewhere. One temptation is to impose a somewhat monolithic order upon available data and findings by relating them to a particular theoretical perspective. As it happens, available over-views of industrial sociology tend very much to be organized in relation to structural functional theories. In *Sociology and Industrial Life* I have attempted to give a judicious critical exposition of a diversity of theoretical approaches. To some extent I have assumed that students will not come to industrial sociology devoid of any sociological knowledge and that they will typically have already been studying some of the general issues of theory and method of sociology. My purpose is to illustrate in an introductory manner the ways in which sociological analysis has been applied to throwing light upon the nature, place and relation of industrial life to society as a whole. I should emphasize that in this brief study I have not attempted to 'cover the ground' to use a phrase made sacred by syllabus devisers. Attempts to do so in my experience either end in mammoth text-books or in something which resembles an annotated bibliography. Yet there is a strategy and a conviction which has conditioned the presentation of this book and I want to articulate it.

The conviction is that in introducing a sociological 'specialism' there is a great deal to be gained from direct engagement in a critical discussion of a number of issues, themes and theories. This is preferable in my opinion to simply presenting a large batch of research findings on work groups, management structures, collective bargaining procedures and so on. The significance of such findings is not only to be evaluated by reference to the soundness of the research techniques, but also by reference to the theoretical perspectives they embrace (implicitly or explicitly) and the contribution they make to on-going debates on particular issues or themes.

Part of the task therefore is to show some of the ways in which sociologists have departed from common-sense discussions about the problems of industrial life (although they have by no means abandoned everyday pragmatic concepts like efficiency, effort, productivity). But what still remains is to convey a sense of the problems which beset the sociologist as he seeks to explain social reality by the invention and skilful development of analytical and often abstract concepts. Some of the things I have in mind here are approached in Part I of the book. Why do some sociologists utilize systems analysis? What elements do they build into the system? How does this affect the utilization of empirical data? Is this to be regarded as something different from or complementary to a social action perspective? How does the work conducted from these various standpoints add to our awareness of the realities of industrial life? In doing this kind of thing, therefore, we do not regard it as sufficient to show why common-sense interpretations of the social world may be found wanting, but also to scrutinize the assumptions and alternatives which sociological interpreters themselves propound and for which certain claims are made from time to time.

I have also taken the view that a critical discussion of this 'specialism' would be very much beside the point without a considered reference back to the classical sociologists. This is in defiance of A. N. Whitehead's well-known dictum that 'a science which hesitates to forget its founders is lost'. We are not so far removed in time from the founders or from the practical and theoretical problems they confronted that we can afford to forget them. Neither for that matter is it at all clear that sociologists have assimilated all that the founding fathers have to teach. But in any case more recent work can only be fully appreciated in the light of classical sociology because much of it proceeds by trying to modify, develop, synthesise or refute the founding fathers. There is a continuity of interest and debate. This I have sought to illustrate in Part II by exploring the anomie theme—the Durkheimian tradition—and in Part III by discussing the alienation theme—the Marxist tradition. In both cases I begin with expository comment on Durkheim and Marx respectively and then proceed to discuss more recent theoretical and empirical work. If students are stimulated to explore some of the questions and issues which are raised, more exhaustively than one could hope to achieve in an introductory book of this kind, then the strategy adopted here will have served its purpose.

Part One

*Industry and Society:
Theoretical Perspectives*

In general terms, one might say it is characteristic of . . . scientific as distinct from non-scientific forms of solving problems that, in the acquisition of knowledge, questions emerge and are solved as a result of an uninterrupted two-way traffic between two layers of knowledge: that of general ideas, theories or models and that of observations or perceptions of specific events.[1]

The main concern of this first chapter is to consider the kinds of problems sociologists have raised in their studies of industrial life. In selecting some for illustrative purposes, the attempt will be made to indicate some of the differences which exist in the character of the explanations offered and the theoretical assumptions within which they are embedded.

(1) *Industry and Society: some developmental themes*

The themes we have chosen to note and comment upon have been selected to draw attention to some of the issues of large-scale social change with which sociologists have been concerned. They are all in a sense a response to the question: is there a logic of industrialism?

Those who want to argue that there is a logic of industrialism essentially want to point to the similarities of structure and process which accompany industrialization in different societies.

Inkeles, for example, has argued that 'insofar as industrialization,

[1] Norbert Elias, 'Problems of Involvement and Detachment' in *BJS*, VII, 1956, 241.

urbanization the development of large-scale bureaucratic structures and their usual accompaniments, create a standard environment with standard institutional pressures for particular groups, to that degree should they produce relatively standard patterns of experience, attitude and value—standard, not uniform pressures.'[1] On the basis of a secondary analysis of cross-cultural surveys, Inkeles seeks to document this claim. For example, a survey of job satisfaction in six industrial societies [2] reveals a similarity in response as between countries, in the sense that those at the top are more satisfied than those at the bottom of the hierarchy and this is broadly reflected as a trend through the standard occupational hierarchy.[3] There are also differences, most notably perhaps, in the fact that the average level of satisfaction as between whole populations differed considerably. Eighty per cent of the United States population sampled expressed attitudes of job satisfaction, twice as many as the German population. (We are not here concerned to argue the significance of these differences, either in methodological or theoretical terms.) Inkeles' point is that when the occupational hierarchy as a whole is taken as the unit of analysis, a uniform pattern concerning job satisfaction can be located. The rank order of an occupational hierarchy as rated by whole industrial populations is, incidentally, remarkably similar as between countries.[4] The thesis which Inkeles and Rossi advanced in that connection was that 'a great deal of weight must be given to the cross-national similarities in social structure which arise from the industrial system and from other common structural factors such as the national state.'[5]

In other words, Inkeles is implying that there are certain evaluations common to industrial populations as a whole which may be cross-culturally validated. But at the same time there are certain cross-cultural comparisons which point to shared perceptions and attitudes of specified occupational groups, which are differentiated from other groups in their own society. Inkeles

[1] Alex Inkeles, 'Industrial Man: the Relation of Status to Experience, Perception and Value' in Henry A. Landsberger (ed.), *Comparative Perspectives on Formal Organisations* (Little, Brown & Co., New York, 1970).
[2] USSR, USA, Germany, Italy, Sweden, Norway.
[3] See Table 1.
[4] See A. Inkeles and P. Rossi, 'National Comparisons of Occupational Prestige' in *AJS*, LXI, 1956. The countries included were Germany, Great Britain, Japan, New Zealand, the USSR and the USA.
[5] *ibid.*, p. 339.

TABLE 1. *National Comparisons of Job Satisfaction, by Occupation*

PERCENTAGE SATISFIED*

USSR		US		Germany	
		Large business	100		
		Small business	91		
Administrative,		Professional	82	Professional	75
professional	77				
Semiprofessional	70			Upper white collar	65
White collar	60	White collar	82	Civil servants	51
				Lower white collar	33
Skilled worker	62	Skilled manual	84	Skilled worker	47
Semiskilled	45	Semiskilled	76	Semiskilled	21
Unskilled	23	Unskilled	72	Unskilled	11
Peasant	12			Farm labor	23

PERCENTAGE SATISFIED*

Italy		Sweden		Norway	
		Upper class	84	Upper class	93
		Middle class	72	Middle Class	88
Skilled worker	68				
Artisan	62	Working class	69	Working class	83
Unskilled	57				
Farm labor	43				

* USSR—percentage answering 'Yes' to: Did you like the job you held in 1940? (Soviet refugee data, Russian Research Center, Harvard University). US—percentage answering 'Yes' to: 'Are you satisfied or dissatisfied with your present job?' (Richard Centers, 'Motivational Aspects of Occupational Stratification', *Journal of Social Psychology*, XXVII [1948], 100). *Germany*—percentage who would choose present occupation in response to: 'If you were again 15 years old and could start again, would you choose your present occupation or another one?' (from German poll data, courtesy of S. M. Lipset). *Italy*—those 'satisfied' or 'fairly satisfied' with work (*Doxa Bolletino*). *Sweden and Norway*—percentage giving 'satisfied' in response to question: 'Are you satisfied with your present occupation, or do you think that something else would suit you better?' (Hadley W. Cantril [ed.], *Public Opinion*, 1935-1946 [Princeton, N.J.: Princeton University Press, 1951], p. 535).

Source: *Comparative Perspectives on Formal Organisations*, op. cit., 270.

advances the thesis that since attitudes and values are shaped by the networks of social relations in which individuals are enmeshed, together with their concomitant rewards and punishments, this helps to explain differential responses within a society. But he is also suggesting that industrialism patterns those differential responses giving rise to cross-cultural similarities. He is, however, prepared to speculate in a way which takes him far beyond his data. If the conditions of life and work with its pattern of rewards and punishments come to be more alike then shared perceptions, attitudes and values will become similar. The implied end is that of a homogenized culture. Inkeles appears to believe that this process is already operative in the USA and that it might come to prevail in other industrial societies:

> Indeed although it seems far off and far-fetched, it could well be that we will in the future come to have a fairly uniform world-culture, in which not only nations but groups within nations will have lost their distinctive sub-cultures. In important respects—exclusive of such elements as language—most people might come to share a uniform, homogeneous culture as citizens of the world. This culture might make them, at least as group members, more or less indistinguishable in perceptual-tendency, opinion and belief not only from their fellow citizens in the same nation and their occupational peers in other nations but from all men everywhere.[1]

This is somewhat unguarded comment, but in such moments the over-view of men's thinking, including that of sociologists, is revealed.

Mention may here be made of two studies which, while they in no way articulate the above speculations, are explicitly derived from Inkeles' perspective. The first is Miller's cross-cultural essay on dockworkers.[2] His major contention is that there are certain widely prevalent conditions associated with dockwork which produce a universal dockworker sub-culture. The statement is

---

[1] See A. Inkeles and P. Rossi, 'National Comparisons of Occupational Prestige' in *AJS*, LXI, 1956.
[2] Raymond C. Miller, *The Dockworker Sub-culture and Some Problems in Cross-Cultural and Cross-Time Generalities*. Comparative Studies in Society and History, II, No. 3, June 1969.

based upon a reading of official reports of governments, trade unions and employers on dockwork, together with studies undertaken by social scientists of particular situations. The conditions Miller so identifies are: the casual notion of employment; the exceptional arduousness, danger and variability of the work; the lack of an occupationally stratified hierarchy and mobility outlets; the lack of regular association with one employer; continuous contact with foreign goods, seamen and ideas; the necessity of living near the docks; and the shared belief of dockers that members of the wider society consider them a low-status group. These conditions are postulated as creating the dockworker sub-culture which embraces the following characteristics: an extraordinary solidarity and undiffused loyalty to fellow dockers; a suspicion of management and outsiders; militant unionism; the appearance of charismatic leaders from the ranks; a 'liberal' political philosophy but a conservative view of changes in work practices; and a 'casual' frame of mind. This last characteristic may be interpreted favourably in terms of an independent spirit or unfavourably in terms of irresponsibility and lack of work discipline.

The argument simply stated then is that although one might upon close inspection discover cultural differences, 'the movement toward similarity predominates, and as a consequence the resultant occupational sub-cultures share some characteristics that cross national and cultural boundaries'. However, Miller singles out the conditions of casual employment and geographical isolation as key factors promoting the sub-culture and points out that as they begin to disappear in advanced industrial societies, the sub-culture will most likely disappear also, notwithstanding the fact that belief patterns associated with the casual era may linger on following de-casualization. This indeed has been the observed experience of countries in Eastern Europe together with the USA and the UK. One should not, however, deduce from this that all industrializing societies must necessarily produce a dockworker's sub-culture at some stage in their industrial history. Two reasons are advanced by Miller for this: first, and most generally, dockwork is derived from the international commercial-industrial system and this is constantly changing over time; secondly, and as an application of this, new forms of industrialization are not to be regarded as repeat performances of the old. Thus 'even factoring out the cultural and resource environment, India will not industrialize in the same way that the United States did, because a highly industrialized America is part of the world in which the Indians live'.[1] In particular it might be argued that a country may attempt to establish effective decasualization schemes early in its industrial history, and, together with other forms of government planning, this may give a different shape to the occupation of dockwork. The dockworker sub-culture as widely manifested in time and place may not then always emerge. It is not a necessary evolutionary stage which must accompany industrialization.

The second study is Form's cross-cultural study of car workers.[2] Form took four countries: the USA, Italy, Argentina and India, all of which vary markedly between themselves as to the degree of industrialization, but in all cases have a car industry. The main hypothesis of the study was 'that workers employed at automative plants in each of four nations will adapt in similar fashion to all areas of their lives; the factory, union, neighbourhood, community and the wider society. That is, the degree of industrialization of the nation in which the worker lives will not greatly affect his adaptation, since a similar adaptation is expected of workers in all four nations. Thus, the technological requirements of a particular type of industry are seen as pre-eminent.'[3] An interviewing programme was then carried out in four plants which had similar production processes. Among the similarities which emerged according to Form were:

1. Workers from all four countries expressed general satisfaction with their work life. About five-sixths of the total sample showed medium or high satisfaction with the particular kind of work they were doing.
2. Approximately 90 per cent of the workers in each plant believed unions were a necessary feature of industrial relations.[4] At least two-thirds of the workers in each plant evaluated unions favourably.[5]

[1] Raymond C. Miller, *The Dockworker Sub-culture and Some Problems in Cross-Cultural and Cross-Time Generalities*. Comparative Studies in Society and History, II, No. 3, June 1969, pp. 313–14.
[2] William H. Form, 'Occupational and Social Integration of Automobile Workers in Four Countries: A Comparative Study' in *Int. Inst. of Comparative Sociology*. X, nos. 1–2, March and June, 1969.
[3] *ibid.*, pp. 96–7.
[4] The lowest figure was 85·6 per cent in Italy, the remainder were all over 90 per cent.
[5] The Italian figure was slightly under two-thirds—65 per cent.

3. Workers exhibited a relatively high and uniform degree of neighbourhood and community involvement, although this was rather less true of the Indian sample.

Form concludes that while it does not follow that the overall similarities he has noted are caused by employment in the car industry 'one is tempted on rational grounds, to suppose that such a relationship may in fact exist. The conclusion seems clear that the industrial sector does have certain common values which seem to have been accepted fairly quickly, even in countries at lower levels of industrial development. The harshest test of the four cases was India where caste, religion and age represent ancient customary values which should affect responses to emerging industrialism. Yet preliminary evidence points to these factors as having little or no impact on the industrial behaviour and values of automobile workers.'[1]

The conclusion is perhaps not so clear as Form suggests. For example, the conclusion on job satisfaction is not always borne out in other studies. For instance, in the Roper survey cited by Blauner, 69 per cent of the sampled American car workers said that if they could go back to the age of fifteen and start life over again, they would choose a different trade or occupation.[2] Goldthorpe reports that the English car-assembly workers sampled at the Vauxhall plant later appeared to derive little intrinsic satisfaction from their work-tasks, with complaints pre-eminently relating to monotony (69 per cent of the sample).[3]

It can also be argued that what one really wants to know in every case is the significance of statements concerning satisfaction. For example, Chinoy suggests the possibility of a chronology of aspirations which may well affect the degree and kind of satisfaction

---

[1] William H. Form, 'Occupational and Social Integration of Automobile Workers in Four Countries: A Comparative Study' in *Int. Jnl. of Comparative Sociology*. X, nos. 1–2, March and June, 1969, p. 116.
[2] See R. Blauner, *Alienation and Freedom*, University of Chicago Press, 1964, p. 202.
[3] John H. Goldthorpe, 'Attitude and Behaviour of Car Assembly Workers: a Deviant Case and a Theoretical Critique' in *BJS*, XVII, No. 3, September 1966, p. 228. See also Charles R. Walker and Robert H. Guest, *The Man on the Assembly Line*, Harvard University Press, 1952. J. Walker and R. Marriott, 'A Study of some attitudes of factory work' in *Occupational Psychology*, XXV, July 1951.

a worker experiences in the car plant.¹ The same man may adjust his pattern of aspirations regarding job promotion ambitions, or outside interests and ambitions, during his occupational career. This may be related to his marital status and family responsibilities, his seniority in the plant, experience of unemployment, and so on. One might then be able to depict a process by which the worker's aspirations, his hopes and desires come to terms with the reality of working class life. Certainly some key considerations may not easily be tapped by survey methods. Chinoy, for example, argues:

> . . . men cannot spend eight hours per day, forty hours each week, in activity which lacks all but instrumental meaning. They therefore try to find some significance in the work they must do. Workers may take pride, for example, in executing skilfully even the routine tasks to which they are assigned.²

The question which remains for Chinoy is in what sense this is to be regarded as positive or real satisfaction. One does not necessarily have to agree with Chinoy's conclusion that real satisfaction cannot be obtained in a context where the central concern of the individual is his economic advancement (within structurally defined limits), to appreciate the nature of the problem for sociological analysis.³ A similar criticism may be offered concerning Form's discussion of attitudes to unionization. He himself notes that there are enormous differences in union ideologies, structures and functions and it is difficult to see how this can be divorced from the assessment. This might, for example, help to account for an otherwise unexplained anomaly, namely that the Argentinian sample recorded the highest percentage of workers with little or no interest in unions (49·5 per cent)⁴ whilst at the same time 60 per cent of the sample were given a high rating for union participation and this was a markedly higher percentage than the other countries.⁵

One of the firmest statements which emphasizes the growing similarities of industrial societies is located in the work of Clark

---

[1] See E. Chinoy, *The Automobile Worker and the American Dream*, Doubleday, 1955.
[2] *ibid.*, pp. 130–1.   [3] On this see also below, pp. 127–128.
[4] The comparable figures for the other countries were: USA 33·7 per cent, Italy 37·6 per cent, and India 30·2 per cent.
[5] The comparable figures for the other countries were: USA 32·9 per cent, Italy 2 per cent, and India 37 per cent.

Kerr and his associates.[1] In *Industrialism and Industrial Man*, for example, developmental questions are explicitly raised in relation to the impact of industry on social life. Does industrialism have an inner logic? What are the inherent tendencies of the industrialization process? What impact do they necessarily have upon workers, managers and governments? Do industrializing societies tend to become more similar to each other or do they retain the variations of their pre-industrial background or develop new diversities?

Kerr refers to what he terms the intrinsic imperatives of industrialism. These appear to be derived from the character and concomitant demands of industrial technology and production. Included here are: the need for a high rate of social and geographical mobility; an educational system functionally suited to supply the technical and professional skills required by technology; a complex and highly differentiated division of labour organized hierarchically and with individuals paid (rewarded, compensated) in relation to their hierarchical position; metropolitan areas as the basis of industrial life with the diminution of cultural variety in society as a result of the growth of transportation and mass communication; an increasing role for governments in regulating and responding to a vast range of problems and needs created by advanced technology.

The industrial society, it is said, develops a distinctive consensus of ideas, beliefs and value judgments. These values typically include a high regard for science and technology and those who practise in that sphere, and a welcoming attitude to technical change. The role of the intellectual in an industrial society is portrayed by Kerr as carrying out 'the function of making explicit a consensus, and of combining discrete beliefs and convictions into a reasonably consistent body of ideas'.[2] Kerr writes in the abstract of a 'pure industrial society' and postulates that such a society is empirically approximated eventually by the inherent features of industrialization.

The place the society starts from and the route it follows are likely to affect its industrial features for many years, but all

[1] See particularly Clark Kerr, J. T. Dunlop, F. H. Harbison and C. A. Myers, *Industrialism and Industrial Man*, Heinemann, London, 1962, and also Clark Kerr, *Labour and Management in Industrial Society*, Doubleday, 1964.
[2] Clark Kerr *et al.*, op. cit., p. 44.

industrialising societies respond to the inherent logic of industrialism itself. The empire of industrialism will embrace the whole world; and such similarities as it decrees will penetrate the outermost points of its sphere of influence, and its sphere comes to be universal. Not one, however, but several roads lead into this new and ultimate empire.[1]

Herbert Spencer himself could not have put it more magisterially. Although diversity of cultural life is recognized, Kerr maintains that such diversity will be eroded by the more powerful influences for cultural uniformity. The picture painted is one of the end of ideological conflicts to be replaced by the age of 'realistic', 'pragmatic' consensus. 'Industrial man is seldom faced with real ideological alternatives within a society. . . . The negotiator takes the place of the prophet. . . . Industrial society must be administered; and the administrators become increasingly benevolent and increasingly skilled. . . . The benevolent political bureaucracy and the benevolent economic oligarchy are matched with the tolerant mass.'[2] And the general direction in which industrial societies are necessarily moving according to Kerr is towards pluralistic industrialism. What is here envisaged is the state as a powerful co-ordinating agency managing conflict between interest groups, establishing and where necessary interpreting or enforcing the rules of the game. Nevertheless the bureaucratic public or private enterprise is described as the dominant institution. This may be subject to checks by the state and also by influential professional and occupational associations. There is much here to remind one of Durkheim, whom we discuss below:[2] but Durkheim does not espouse a doctrine of historical inevitability, and for him the issue of social inequality is more problematical. The Marxist vision of a classless society is presented, but with the state still very much present: 'Class warfare will be forgotten and in its place will be the bureaucratic contest of interest group against interest group . . . memos will flow instead of blood.'[4]

It must be said at once that it is difficult to test an immanent evolutionary theory of this sort. One may draw attention to various forms of evidence which it might be imagined would

---

[1] Clark Kerr et al., op. cit., p. 46.   [2] ibid., p. 283.   [3] See pp. 73-91.
[4] ibid., p. 292.

cast doubt upon its validity—for example, comparative and longitudinal information on, say, income distribution or social mobility in industrial societies. This, indeed, Goldthorpe has done,[1] but it is always possible to re-interpret such evidence as a hindrance to the ultimate inevitable long-term development. What one is really being offered, however, is a statement about the necessary development of industrial societies based on a functionalist view of society which sees the end of ideology as both inevitable and a mark of the just society. It is above all to reify the concept of society and it is most notable that the action of man is left to count for nothing in technological utopia. Yet of course if one starts one's analysis with man as actor and contends for the possibility of political action as a decisive element affecting the direction in which particular industrial societies may move, then a far more open-ended prospect is envisaged. It might make the sociologist's job of prediction more difficult, but at least he will not have succumbed to the sorcery of historicism. (One cannot refrain however from observing the curious irony that Popper, the scourge of historicists and the advocate of social engineering is, in his advocacy, not unlike the intellectuals to whom the historicist Kerr assigns the task of articulating the pragmatic consensus of industrial society.) In place of technological imperatives we would suggest it is more appropriate to speak of technological constraints. It is man's activity in relation to those constraints that has to be explored: man—as conscious, reasoning and reflective being—who, in and through his relation with other men, creates and re-creates the social world.

Feldman and Moore have offered a more cautious developmental thesis when considering the impact of industry on society.[2] They argue that industrial societies share a core set of social structures: notably the factory system of production and market system, which permits an extensive commercialization of goods and services, a stratification system based on an extensive and complex division of labour, and an education system which is able to filter

[1] John H. Goldthorpe, 'Social Stratification in Industrial Society' in *The Sociological Review Monograph*, No. 8, 'The Development of Industrial Societies' October 1964.
[2] See Arnold S. Feldman and Wilbert E. Moore, 'Industrialisation and Industrialism: Convergence and Differentiation' in W. A. Faunce and W. H. Form (eds.), *Comparative Perspectives on Industrial Society*, Little, Brown & Co., New York, 1969.

individuals adequately into the occupational and stratification system.

Although Kerr writes of the different routes that may be followed in reaching 'pure industrial society', for Feldman and Moore these different routes or trajectories as they term them, are of great importance. For example, the historical era in which industrial change takes place may affect the targets which political and economic agencies in a society aim for. Assuredly twentieth-century industrializing nations do not aim for the same targets as nineteenth-century England did. The fact that there are observable differences in the rates and character of change in industrializing societies is not something to be shrugged off as temporary, prior to a final convergence. This is to over-emphasize the degree of functional interdependence necessary for industrial societies to exist. Indeed given Moore's known sympathy for functionalist theory, the author's eventual emphasis is all the more interesting:

> ... in order to predict the future course of industrial societies less is likely to be gained by fitting out an equilibrium model—the kinds of structures that are functionally consistent with industrialism but 'still' incompletely developed—than by direct attention to the sources of continuous change.... The dynamic features of industrial societies and the reality of their relative autonomy as political entities leads us to expect actually increasing differentiation in their specific tensions and strains and in the way these are partially controlled.[1]

It is notably the recognition of the role of political activity in affecting the direction of particular industrial societies that decisively separates Feldman and Moore from Kerr's position.

[1] op. cit., p. 68.

(2) Systems Analysis and Industrial Behaviour

The very widespread use of what can generally be termed 'systems analysis' as a framework in which explanations of many facets of industrial life are offered, cannot go unmentioned. It is noteworthy that two of the most widely used American text-books of industrial sociology are explicitly organized in this way.[1] Schneider, for example, operates within a Parsonian framework. For him the structural-functional approach is treated as a form of analysis which relates the various roles, groups, institutions and personalities in a social system to the needs of the social system as a whole. But what are these needs or functional prerequisites? Schneider concedes that no single answer can be given because social systems differ widely in their nature, but he does claim that there are certain universal prerequisites for the existence and stability of a social system such as the maintenance of order, the motivation of actors within the system, communication between the actors, a consensus in beliefs, values and definitions, and protection from the encroachment of external forces. There are of course formidable methodological problems involved in being able to establish empirically when a system is 'stable' or in 'equilibrium'.[2] For our own part we take the view that such an approach quickly comes up against the danger of reification: the social system is given a persona. Thus, typically, we read about how the system must function 'to maintain itself', must secure adequate participation of actors in the system 'in order to persist and develop', must minimize conflict 'in order to accomplish its major purposes', and so on. And often it seems that there is some kind of life-force operating, such that there are 'strains' towards functional integration of the parts of the system, or a necessary process of system adaptation to the environment (variously defined).

[1] D. C. Miller and W. H. Form, *Industrial Sociology. The Sociology of Work Organisations*, Harper & Row, 1964. E. V. Schneider, *Industrial Sociology*, McGraw-Hill, 1969.
[2] See for example, Richard S. Rudner, *Philosophy of Social Science*, Prentice-Hall, 1966, Chapter 5.

26   SOCIOLOGY AND INDUSTRIAL LIFE

Without attempting anything like an exhaustive critique or review of the very considerable literature of systems theory applied to the study of industrial life, we will draw attention to several of the better known variants.

One of the earliest examples of systems analysis in industrial sociology is of course contained in *Management and the Worker*.[1] One construct began with the notion of an individual working 'effectively' and defined, accordingly, as being in a state of

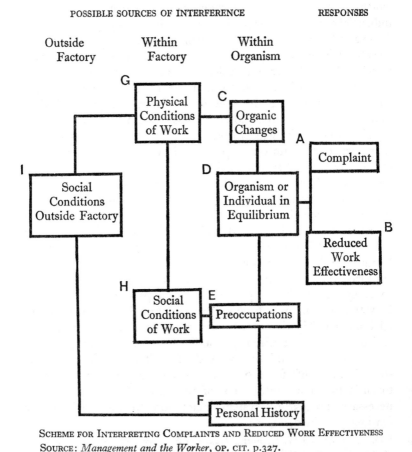

SCHEME FOR INTERPRETING COMPLAINTS AND REDUCED WORK EFFECTIVENESS
SOURCE: *Management and the Worker*, OP. CIT. p.327.

---

[1] F. J. Roethlisberger and W. J. Dickson, *Management and the Worker*, John Wiley & Sons Inc., New York, 1964.

equilibrium. If, however, the individual began to complain about his work situation and/or reduced his work effort it was suggested that this could be explained by looking at possible sources of interference (see diagram). The investigators used this scheme to suggest that in principle disequilibrium might be of various kinds ranging from physiologically based fatigue to social problems at work or outside the factory. The interconnection between the boxes in the diagram is intended to convey mutual influences which make it difficult to disentangle cause and effect. But the researchers did report that 'a very common form of induced unbalance in the worker which diminishes his capacity to work effectively can best be understood in terms in which (*a*) capacity to work, or to fix and sustain attention, is to be regarded as a product of the personal equilibrium of the worker with the social reality about him (see relations I, H, E and D in diagram), and (*b*) any circumstances adversely affecting their personal equilibrium is likely to reverse itself in a reduced capacity for active work in obsessive reveries and in responses' (see relations E, D, A and B in diagram).[1] The argument is also advanced that disequilibrium is a temporary state and will be followed either by the restoration of the old balance or the establishment of a new equilibrium. Perhaps the point to bear in mind here, however, is that work effectiveness/individual equilibrium relationship tends to be interpreted in managerial terms—the task of the social science consultant is how to advise on the maintenance or re-creation of a favourable equilibrium. It does not follow, as we seek to show below,[2] that 'reduced worker effectiveness' has to be understood as an 'irrational' response. There is the further difficulty of deciding what constitutes 'temporary disequilibrium'. Given conflicts of definition, for example, over what constitutes a 'fair' wage or a 'fair' day's work, disequilibrium may be the more characteristic and permanent phenomenon.

Roethlisberger also seeks to explain employee satisfaction or dissatisfaction in system terms with the employee's position or status and his own attitude to it as crucial (see diagram). What is important about this analysis is that it does take seriously the individual's own definition of his situation. On the one hand the

[1] F. J. Roethlisberger and W. J. Dickson, *Management and the Worker*, John Wiley & Sons Inc., New York, 1964, pp. 327-8.
[2] See pp. 45-64.

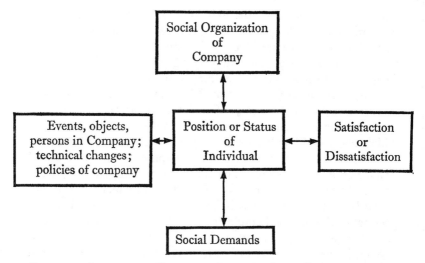

SCHEME FOR INTERPRETING COMPLAINTS INVOLVING SOCIAL INTERRELATIONSHIPS OF EMPLOYEES
SOURCE: *Management and the Worker*, OP. CIT., p.375.

organization is seen as the vehicle of a system of beliefs and practices. A premium, for example, might be placed on such values as 'efficiency', or 'service'. Roethlisberger and Dickson then observe:

> In these terms it is then possible to understand the effect upon the individual of—or the meanings assigned by the individual to—the events, objects and features of his environment, such as hours of work, wages, etc. Only then is it possible to see what effect changes in the working environment have upon the social reorganization to which the employee has become accustomed, or upon that ideal type of equilibrium which he desires.[1]

This is an important move away from a reified systems approach, and helps us to recognize that the systems perspective and the social action perspective (which we discuss below) can move together insofar as the first takes account of the actor as definer of, and agent in, his situation, and the latter comes to terms with social structure as constraint on action.

[1] *Management and the Worker*, p. 375.

It is also worth recognizing that the systems analyses put forward in the Hawthorne studies do not treat the plant as a 'closed' system. Not only is this reflected in the diagram, but Roethlisberger and Dickson explicitly argue the point:

> . . . the relation of the individual employee to the company is not a closed system. All the values of the individual cannot be accounted for by the social organization of the company. The meaning a person assigns to his position depends on whether or not that position is allowing him to fulfill the social demands he is making of his work. The attitude and significance of his work is not defined so much by his relation to the company as by his relation to the wider social reality. Only in terms of this latter relation can the different attitudes of satisfaction or dissatisfaction of individuals who are presumably enjoying the same working environment and occupational status be understood.[1]

Homans in his study *The Human Group*,[2] and partly derived from his own association with the Hawthorne investigations, offered another form of systems analysis. Essentially he employed three concepts—activity, sentiment and interaction—as constituting the interdependent elements of the system. Activities define the things men do, interactions the human contacts they establish, and sentiments the attitudes and opinions they express. Such a system is seen as generating its own norms of behaviour which, of course, in turn come to affect the way the system operates. There is, in other words, a feed-back mechanism which, other things being equal, serves to maintain the system or to enable it to survive in what might be a threatening environment. If Homans had chosen to restrict his discussion to a closed system he would have, in his language, simply written about the internal system. However, he recognizes that this would be very unrealistic and therefore uses the term external system to denote the fact that activities, sentiments and interactions can be conditioned by the environment. (The term environment is, for him, to be classified into three further interrelated categories: physical, social and technical.) So, for example, when he discusses the Bank Wiring Observation Room study at Hawthorne, he offers a

[1] *Management and the Worker*, pp. 375–6.
[2] George Homans, *The Human Group*, Routledge & Kegan Paul, 1950.

comment which is of theoretical importance and relevance when considering further research strategies:

> Many of the ideas about restriction of output must have been picked up from (fellow workers outside the group). Important also was the Chicago of the early years of the depression, and so were the groups in which men participated outside of their work. Their membership in families certainly had a direct effect on their behaviour, particularly on their motives for work, and so did their looser membership in neighbourhoods, social classes and churches. We mention these aspects of the environment, not because we know enough about their effect on the wiremen, but because they ought to be looked at in the study of a group. We are not just analysing the environment of the Bank Wiring Observation Room but setting up a check list for future use.[1]

Some attempt has been made to follow this up, notably in the work of Lupton[2] and Cunnison.[3] The basic question which they consider is: what social factors determine the establishment of production norms in factories? They both see levels of output and earnings as being the outcome of the interplay of a complex of factors, some of which lie outside the workshop itself and, therefore, outside the power of management or workers to control. In the presentation of their case studies, in Garment Manufacture and Electrical Engineering, the factors that are considered in practice may be summarized in the following way:

External factors
(1) Market situation
    (a) its stability
    (b) its size
(2) Relations with competitors
(3) Location of industry
(4) Trade Union organization
    (a) National level
    (b) Local level
(5) Type of product.

Internal factors
(1) Method of wage payment
(2) Nature of the productive system
(3) Sex of workers
(4) Workshop social structure
(5) Degree of congruence and conflict in management-worker relationship
(6) Ratio of labour costs to total costs.

---

[1] George Homans, *The Human Group*, Routledge & Kegan Paul, 1950, pp. 89–90.
[2] T. Lupton, *On the Shop Floor*, Pergamon Press, Oxford, 1963.
[3] Sheila Cunnison, *Wages and Work Allocation: a Study of Social Relations in a Garment Workshop*, Tavistock Publications, 1963.

Since the work of Lupton and Cunnison has been done by participant observation techniques in the factory, the main emphasis has in fact been on the internal factors so far as detailed analysis is concerned. In a critique of their own work,[1] they admit that their treatment of external factors 'has been ex post facto as well as ad hoc'.[2] This should not detract from the potential power of the theoretical approach as revealed in the following statement:

> Although it is not possible, with present knowledge, to assess the relative significance of the various external factors, we nevertheless feel justified in advancing the hypothesis that in industries characterized by clusters of factors such as a high degree of competition, easy entry of new units of production, high labour costs, and weak unionism, collective control over output by workers is less likely to be found than in industries where the reverse is true. We would also say that these clusters of characteristics are more commonly found in some industries than others.[3]

Lupton and Cunnison also explicitly comment on the fact that to discuss the linkage of the workshop with the environment demands that one makes an analytical distinction between overlapping and inclusive social systems. The first consideration embodies the notion that 'there are segments of other social systems which are latent in a single workshop. For instance, a man's role as a father, or as a member of a social class, may affect his behaviour in the workshop.'[4] Such considerations, for example, may help to explain the behaviour of workers who deviate from established workshop norms. The workshop may also be actively reckoned as included in a wider social system: for example, the exchange system of the market. In this way the economic structure of the industry can be influenced, and this may be reflected, for example, in the productive system of the workshop, which itself operates as a constraint on workshop social structure. What is positively suggested, therefore, is the need for a more rigorous treatment of external factors. The participant observer's know-

---

[1] T. Lupton and Sheila Cunnison, 'Workshop Behaviour' in M. Gluckman (ed.), *Closed Systems and Open Minds*, Oliver and Boyd, 1964.
[2] *ibid.*, p. 146.   [3] *ibid.*, p. 124.   [4] *ibid.*, p. 125.

ledge of factory life will have to be supplemented by a more sophisticated understanding of the economic system of which the factory is a part, and a thorough exploration of the overlapping systems, undertaken almost inevitably by a team of sociologists rather than one individual researcher.

In order to obtain information about overlapping systems, Lupton and Cunnison suggest that this may well involve 'intensive interviews with the families of workers . . . to map out kinship networks, to discover financial obligations to kin, and the present occupation, educational and occupational histories, and leisure time activities of the worker and his family'.[1] The interesting point about this is that it begins to look very much like the kind of approach advocated by Goldthorpe[2] (the social action approach which is discussed below) even though it is arrived at through a social systems style of conceptualization.

It would be too cynical to take the view that systems analyses have tended to dominate the study of industrial life because they hold out the possibility of managerial control over hitherto unpredictable factors, including the labour force. It is nonetheless true that many systems studies are preoccupied with factors affecting worker productivity or managerial efficiency. And the social science consultancy approach does appear sometimes to hold out hopes for creating a more 'healthy', 'normal' or 'better adapted' system. It is as though one seeks to understand 'the system' in order to control it (notwithstanding its immanent tendencies) on one's own terms. The research consultancy work of the Tavistock Institute provides some notable examples of this approach. There is no doubt about a fundamental assumption in the Glacier Metal Project, namely that the worker-management relations could be restructured to create a 'healthier' factory culture. The development of a joint consultation system was advocated, at least in part, for its therapeutic value. The concentration of the study on the internal structure of the factory resulted in a very muted treatment of environmental factors. One suspects, in particular, that the trade unions as social organizations claiming worker allegiance and providing countervailing

[1] T. Lupton and Sheila Cunnison, 'Workshop Behaviour' in M. Gluckman (ed.), *Closed Systems and Open Minds*, Oliver and Boyd, 1964, p. 125.
[2] John H. Goldthorpe, 'Attitudes and Behaviour of Car Assembly Workers: a Deviant Case and a Theoretical Critique', in *B.J.S.*, XVII, No. 3, September 1966.

sources of power and authority are treated with some ambivalence by Jacques.[1] Certainly they could threaten the planned stability of the factory's social system.

One organizing concept which has come to be used increasingly by Tavistock researchers is the 'socio-technical system'. For blue-collar workers this has been particularly applied to studies of the mining industry in the UK.[2] The socio-technical system is described as having social, technological and economic dimensions, 'all of which are interdependent, but all of which have independent values of their own'.[3] However, even in their major study, one finds that the focus is on the socio-psychological system. Because of this, it is not possible—in my judgment—for Trist et al. to give empirical meaning to the statement: 'the optimization of the whole (socio-technical system) tends to require a less than optimum state for each separate dimension'.[4] One may further note that within the socio-psychological system the organizational level selected for study is the primary work-group considered in the context of the surrounding 'seam society'. There is a further limitation that, although one is invited to consider the research as based upon an 'open socio-technical systems' approach in order that the enterprise may be recognized as interacting with its environment, the environment is effectively defined as the 'seam society'. The local community, which is so fruitfully discussed in Dennis et al.[5] is scarcely mentioned, and the role of the N.U.M. is not extensively discussed, although the importance of the miners' lodge as a local negotiating unit is recognized. Trist and his colleagues observe that 'in traditional mining methods, control and regulation of work at the coal face were carried out autonomously by the working group, which developed customs of self-regulation, task continuity and role rotation appropriate to the underground situation. Seam officials provided services to the independently producing work places and, because of the slow tempo of production, had little difficulty in co-ordinating operations in the seam as a whole.'[6] But can the responsible autonomy of the work group be realized in more

[1] E. Jacques, *The Changing Culture of a Factory*, Tavistock Publications, 1951.
[2] Especially: E. L. Trist et al., *Organisational Choice*, Tavistock Publications, 1963.
[3] *ibid.*, p. 6.   [4] *ibid.*, p. 7.
[5] N. Dennis et al., *Coal is our Life*, Eyre and Spottiswoode, 1950.
[6] Trist et al., op. cit., p. 29.

fully mechanized pits? The researchers claim that it can, despite the fact that primary work groups in this context may include a membership of up to fifty: a finding which challenges much small group theory in social psychology. They argue that since this releases management from the burden of close supervision, allowing them to concentrate on 'providing effective communications within the seam, on anticipating the support required for operations at the face, and on giving attention to longer-term planning and development'[1] higher productivity ensues: a claim which runs counter to classical scientific management precepts. Although one may wish to question some of the theoretical scaffolding of the study, the reminder of the validity of organizational choice (and the differing consequences which flow from these choices) within a given technological and economic context, provides a cautionary note to those who adopt a stance of technological determinism in discussing work group behaviour.

Turning now to the management side of behaviour, Miller and Rice have recently documented a number of interesting case studies. We will take as an illustration their work in the steel industry.[1] The book explores the proposition that 'any enterprise requires three forms of organization—the first to control task performance: the second to ensure the commitment of its members to enterprise objectives: and the third to regulate relations between task and sentient systems.'[2] The sentient system is defined as that system or group that demands and receives loyalty from its members.

In earlier work the Tavistock researchers had tended to suggest that the goals of organizational design should be to permit effective task performance and satisfy human needs by ensuring that the boundaries of task systems and sentient systems coincide. In *Systems of Organisation* they no longer say this. Indeed, they argue very differently:

> ... the organization in which it is possible to match sentient groups to task—and so make task and sentient groups coincide—are the exception rather than the rule. What is more, a group

[1] Trist *et al.*, op. cit., p. 295.
[2] E. J. Miller and A. K. Rice, *Systems of Organization*, Tavistock Publications, 1967.
[3] *ibid.*, xiii.

that shares its sentient boundary with that of an activity system is all too likely to become committed to that particular system so that, although both efficiency and satisfaction may be greater in the short run, in the long run such an organization is likely to inhibit technical change. Unconsciously, the group may come to redefine its primary task and behave as if this had become the defence of an obsolescent system. The group then resists, irrationally and vehemently, any changes in the activities of the task system that might disturb established roles and relationships.[1]

With this in mind, we may note that Miller and Rice discuss the organizational problems involved in building a new steel works on a 'green field site'. In an interesting account the researchers discuss why certain tendencies towards managerial conservatism could be detected. For example, an innovating decision was made by the General Manager (Operational) to bring operational management in at a much earlier stage in plant construction than is common. But the basis upon which they were recruited was an organization chart prepared by the organization and methods department looking suspiciously like the management structure of other plants in the company. This gave rise in practice to a paradox:

> As organization designers, they were expected to develop management philosophies, techniques, methods, policies and procedures that would take advantage of the green-field situation. But it was not foreseen that in carrying out the task of organization design they might well call into question the conventional assumptions that underlay the chart by which they had been appointed. In other words, either the organization for operations would act as a strain on organizational design, or, alternatively, the organization-design activities would threaten the operation organization.[2]

A strong desire was noted among the operating managers to retreat into familiar roles based on past experiences and this

[1] E. J. Miller and A. K. Rice, *Systems of Organization*, Tavistock Publications, 1967, p. 31.
[2] *ibid.*, p. 148.

inhibited organizational design changes that might otherwise have been made.

The other major point made was that the engineering and services specialist managers, who had considerable executive powers in the setting-up phase of the plant, tried to prolong that power and extend it into the operating period. This, it was said, added to the uncertainty of the production managers who, having been brought in at this earlier stage, were, in effect, trying to impose the future on to the present. What seems to be implied is that a power struggle between these 'sentient groups' was going on, which in the researchers' view hindered the finding of successful solutions to task requirements in the here and now. That there was such a struggle is scarcely surprising to a sociologist. In conditions of innovation and uncertainty, who could definitively say when 'production' should take over and 'engineers' give up executive authority? Some might argue that engineers could effectively take over in this kind of new plant. But in any case, managers, like nature, abhor a vacuum and struggle with one another to fill it. Miller and Rice argue, however, that the conventional management structure which had been imported was in some respects anti-innovation and, in this sense, created a number of unnecessary problems: 'Innovations appeared to be perceived as threatening predetermined patterns of power and status in the future operating organization. These production managers sometimes defended their future operating roles against change instead of designing change into the operating system.'[1] Miller and Rice argue, therefore, that organizational design teams should not occupy or be promised roles in operating systems because this 'contaminates' their design prescriptions. Their commitment should rather be exclusively to design work. The implication appears to be that organizational designers are commando groups moving from one situation to another in industry. One senses the practical objections of a general manager at losing men with valued experience. But, as Miller and Rice point out, one can always use outside consultants in helping to solve design problems!

The above points notwithstanding, top managerial initiative could, and did, give rise to key decisions which led in practice to

[1] E. J. Miller and A. K. Rice, *Systems of Organization*, Tavistock Publications, 1967, pp. 155-6

changes in the effective management of the plant. Decisions, for example, to introduce control systems in the spheres of production, quality, budget and costs, manning and wages, and plant maintenance; and, so far as possible, to introduce process automation; were of considerable significance in their effects. In particular, the interlinked aspects of power, status and autonomy of decision making were diminished for the production manager. Centralized information systems made their work more speedily open to top management appraisals. The implementation of these systems also enhanced the power of the specialist and engineering managers instituting and running them. Production work in general, and the production manager's role in particular, was, so to speak, 'demystified'. This was well illustrated by the fact that after the plant had been operational for a year, a departmental manager was appointed with no technical knowledge of the production process. His success, which in the traditional context of steelmaking, would have been unbelievable, was attributed by the researchers to the fact that the predominant task of the departmental manager had in fact become 'to operate the control systems in such a way as to meet the various standards laid down'.[1]

Broadly, the Tavistock researchers recommend greater flexibility of thinking both in the creation of appropriate and the restructuring of inappropriate management organization. The definition of what is entailed in appropriate organization is reflected in their concept of 'strategy-mix'. If the primary task of industrial enterprise is to maximize its return from its total transactions with its environment, the appropriate strategy-mix takes account of environment and technological opportunities and constraints in order to realize the primary task.

Situation	*Dominant strategy*
Buyers' market	Maximize customer satisfaction
Sellers' market	
Low activity market or capital invested in anticipation of future growth in demand	Maximize production Conserve resources

[1] E. J. Miller and A. K. Rice *Systems of Organization*, Tavistock Publications, 1967.

The chart above summarizes the strategy defined as appropriate for particular situations, with the provision that they cannot be mutually exclusive. 'What is required is an optimum strategy-mix which assigns the appropriate weighting to each component strategy in the light of prevailing circumstances.'[1]

Certainly, it could be plausibly argued that in the steel context maximization of production was too commonly treated as the 'obvious' strategy and that this broadly determined the conventional management structures found in the industry. At the same time, the adoption of an 'appropriate' strategy depends in the first place on the availability and agreed interpretation of a wide range of economic and technological facts and trends. One should not oversimplify what is really involved here. There is a problem of the knowledge gap even if 'the light of prevailing circumstances' is treated in static terms. Given a complex market situation and a sophisticated technology, identifying 'correct' managerial behaviour, and hence the appropriate managerial structure, is no easy matter. The problem is further compounded, of course, as soon as one recognizes that the real managerial problem is a dynamic one: understanding not only the nature of the present market situation and responding to it, but also identifying the direction of change in the market and evaluating company strategy in the light of changing circumstances.

In the second place, there are what might be called normative bounds to particular strategies in which account is taken of the likely social as well as economic consequences of particular policies. Since vesting day, this point takes on extra force in the sense that the nationalized industry explicitly accepted that striving for long-term profitability had always to be placed in the context of social responsibility. So, for example, Ron Smith, the BSC Personnel and Social Policy Director, comments:

> We . . . recognize that the cost of change to communities or individuals can be reduced by our own efforts in, for example, such vital matters as the timing of redundancies in relation to the trade cycle, adequate warning of impending redundancy,

[1] E. J. Miller and D. Armstrong, 'The influence of advanced technology on the structure of management organization' in (J. Stieber (ed.), *Employment problems of automation and advanced technology: an international perspective*, Macmillan, 1964, 332.

additional payments, the provision of retraining and redeployment facilities, collaboration with other enterprises and in influencing Government policy to make special payments to the industry or an appropriate adjustment to the BSC's financial objectives.[1]

Thirdly, one has to recognize the existence of a very formidable constraint on organizational strategy since vesting day: the British Steel Corporation. The Annual Report of the BSC for 1967–8 makes clear that investment proposals for particular plants were to be judged not only by how likely they appeared to promote increased profitability, meet the changing patterns and requirements of customers, and lower operating costs, but also in terms of how consistent they were with the Corporation's long-term plans.

The notion of strategy-mix implies an attempt to meet one's objectives as effectively as possible. It does not take long to realize that this might well be over and against the wishes of others, and so we are confronted with questions of power relations. A consultant may well give advice to a particular group on how to achieve its objectives (and even help it to define its objectives more precisely). To that extent, he becomes identified with the group he advises. The client (which in principle could be a local management group, a trade union, or the BSC) has a 'problem' and will be satisfied when the social scientist has provided him with a 'solution'. But one group's solution is another group's problem. The logical end-product is that each interest group employ its own social scientist to plan strategies and counter-strategies! There would at least be a touch of poetic justice about it. An alternative view is that the relevance of sociological analysis to those in industry is of a different order: 'the sociological problem is not so much why some things "go wrong" from the viewpoint of the authorities and the management of the social scene but how the whole system works in the first place, what are its presuppositions and by what means it is held together.'[2]

[1] R. Smith, *British Steel*, August 1968.
[2] P. L. Berger, *Invitation to Sociology*, Doubleday, New York, 1963, p. 37.

(3) *The Social Action Perspective*

What has come to be called the action frame of reference in sociological analysis is pre-eminently derived from the work of Max Weber. In terms of definition Weber says that 'action is social insofar as by virtue of the subjective meaning attached to it by the acting individual (or individuals) it takes account of the behaviour of others and is thereby oriented in its course'.[1] This orientation to the behaviour of others may relate to their past, present or future expected behaviour. If, of course, one looked at 'subjective' meaning in all its ramifications for each individual, sociology might disappear under a welter of idiosyncrasies and contingent experiences. Such an approach may lead to the production of ethnographical accounts hopefully authentic and probably plausible. A good recent example of such writing in the field of our interest is found in the two-volume collection: *Work*.[2] Personal accounts are given of what it feels like to be say, a policeman, an atomic energy scientist, a toolmaker, and so on. How are their individual lives and experiences shaped by work? What are their hopes and frustrations? There need be no doubt that such writing can be most valuable and full of self-insight. It might be argued that the very act of writing brings a measure of self-consciousness to the work experience which in some respects is artificial. And it cannot of course be assumed that 'the whole story' is reproduced. We may be left wondering also, perhaps, how far the story that is told is in any way typical or representative of others in a similar occupation. This, in principle, is to bring the sociologist back in.

The highly individual approach is also reflected in biographical or ethnological accounts of behaviour when the emphasis is on giving an account of the texture of the actions observed. This has something in common with certain kinds of novel writing: one thinks in our context of Zola's *Germinal* or Tressell's *The Ragged Trousered Philanthropist* as cases in point. It also has something in common with serious journalistic writing. It is worth recalling

---

[1] M. Weber, *The Theory of Social and Economic Organisation* (edited and with an introduction by Talcott Parsons), Free Press, 1964, p. 88.
[2] Ronald Fraser (ed.), *Work* (2 vols.), Penguin, 1968, 1969.

that Robert E. Park who was such an enormous influence on the work of the Chicago school of sociologists went into sociology from journalism with the view that the sociologist should be a kind of super-reporter. His sociological credentials should be his concern for accuracy in reporting, his ability to describe what he sees in a detached way without introducing his own personal moral judgments. The prompting question was always 'what is actually going on?' And the question was answered in terms of the so-called natural history approach, and, amongst other specific interests, led to significant contributions to the study of occupations. An example of this genre is Gold's study of the apartment-building janitor.[1]

In his account Gold describes the content of the job—its responsibilities: how the janitor categorizes tenants into 'good' or 'bad' and tries where possible to change the 'bad' into the 'good': the kind of relationships he establishes with other janitors and how he conceives his career pattern. The essay is interspersed with extended comments from janitors with whom he had met and talked. We learn for example of the ambiguous attitude towards the public which he has. He recognizes that he is there to serve them but they can, by not being 'reasonable', be an endless source of annoyance and interfere with his routine. How does one adjust? A janitor explains:

> The first year as a janitor you are sensitive about a lot of things; then you get hardened to it. You get mad at the tenant if he complains, and if the complaint is legitimate you get mad at yourself. Whenever you get to meet the public you kinda gotta put a shell on you, you know. It's like when you have a position of trust. You're blamed if something is found missing, but you never get credit if extras are found.[2]

In fact, of course, such natural histories are never a product of pure observation—observations are in the nature of the case selective—and there is an element of the observer's interpretation written into the account. There is also the element of inference, conjecture, and anticipation of possible theories. Gold writes for

---

[1] Raymond L. Gold, 'In the Basement—The Apartment-Building Janitor', in Peter L. Berger (ed.), *The Human Shape of Work*, Macmillan, 1964.
[2] *ibid.*, p. 11.

example of the problem of mandate which arises actually in the context of city life when individuals have not agreed clearly on reciprocal rights and obligations. This uncertainty leads to checking and counter-checking of mandates and is seen as a source of moral tension. Such tensions 'pervade all areas of human activity . . . where rapid change and experimentation with life styles foster highly tentative, groping attitudes towards ideas, men, acts and artifacts. My story of the apartment-building janitor illustrates well some of the basic problems of all workers and also of all residents in urbanized areas of our society.'[1] The same form of inference is found in another Chicago study on the cab-drivers.[2] The essay (which typically for the Chicago school included participant observation as a technique) focuses upon the ways in which, given the importance of the tip for the cab-driver's take-home pay, he tries to control the vagaries of tipping. Essentially this is attempted through categorizing the passenger as a certain type, 'sport', 'businessman', 'lady shopper' and so on, and then pursuing the tactics felt to be most appropriate for getting a good tip. Davis interprets much of this as a game and argues that outcomes are, however, taken as a bench mark in the sense that, arising from the transient character of the relationships entered into, the constraints operating in the client-practitioner role performances are very weak indeed. The vagaries of the relationship lead Davis to infer what might happen in other currently more stable client-practitioner situations given certain developments:

> For, given too great and random a circulation of clients among practitioners . . . the danger is that informal social-control net-works would not come into being, and, as in big-city cab-driving, relations between servers and served would become reputationless, anonymous and narrowly calculative.[3]

Not only do these studies offer theoretical conjectures of this order, but such individual studies are used by sociologists as a

---

[1] Raymond L. Gold, 'In the Basement—The Apartment-Building Janitor', in Peter L. Berger (ed.), *The Human Shape of Work*, Macmillan, 1964, p. 48.
[2] Fred Davis, 'The Cabdriver and his Fare: Relationship' in Gerald D. Bell (ed.), *Organisations and Human Behaviour*, Prentice-Hall, 1967.
[3] *ibid.*, p. 271.

THE SOCIAL ACTION PERSPECTIVE

basis for making conditional generalizations or probability statements. Everett Hughes, who supervised many of the Chicago monographs, reported his own growing awareness of what might be cumulatively achieved:

> At first I thought of these studies as merely interesting and informative for what they would tell about people who do these humbler jobs, i.e. in American ethnology. I have now come to the belief that although the problems of people in these lines of work are interesting and important as any other, their deeper value lies in the insights they yield about work behaviour in any and all occupations. It is not that it puts one into a position to debunk the others, but simply that processes which are hidden in other occupations come more readily to view in these lowly ones. We may be dealing here with a fundamental matter of method in social science, that of finding the best possible laboratory for the study of a given series of mechanisms.[1]

If now we return to Weber, it is to recall that at one level he was concerned with 'actual action' and at another with 'types of action'. The second was the explicitly sociological level, but it was not constructed in a vacuum. Rather, one has the impression of a method in which types of action were being distilled (as pure types) from what one had already discovered about 'actual action', and they were then turned again on to social life in order to explain what went on more cogently. One needed, however, the caricature or over-simplification implicit in the type to make any sense at all of explaining the complexity of the real world with all its contingencies and imponderables.

It is in that spirit that we find Weber constructing his types of social action. When, for example, he classifies social action in relation to the mode of orientation, he distinguishes four types:

1. Action oriented to tradition—a habit bound response.
2. Action dominated by the feeling state of the actor—the spontaneous expression of emotion.
3. Action rationally oriented towards an absolute value—the self-conscious formulation of the attitude values governing one's action and the detailed working out of one's behaviour in relation to those values.

[1] E. C. Hughes, *Men and Their Work*, Free Press, 1958, p. 49.

4. Action rationally oriented to a system of discrete individual ends in which the end, the means, and the secondary results are all rationally taken into account and weighed. (Essentially there is implied here a much more developed concern with the consequences of one's actions than in the first three types.)

Weber then suggests that if one discovers 'actual action' which approximates to one or other of these types, then there is evidence that one has a useful sociological tool at one's disposal. Since the possibilities of classification are not necessarily exhausted in such a typology, one may if necessary construct further pure types.

Now it could be argued (with qualifications) that the ethnological style of the Chicago occupational sociologists tends towards a *micro*-sociology, and equally that Weber's classification of types of social action leads in the same direction. The fact that this is not so can best be noted by recognizing that the social action is linked to *macro* considerations in the concept of 'legitimate order'.

> Action, especially social action which involves social relationships, may be oriented by the actors to a *belief* in the existence of a 'legitimate order'. The probability that action will actually empirically be so oriented will be called the *validity* of the order in question.[1]

What Weber then encourages the sociologist to do is first to ask questions about the validity of an order. If it is generally recognized as valid one wants to know what groups conform and what seek to evade the claims of the legitimate order. Further, one may wish to know upon what basis the validity of the order rests, and to what extent competing interpretations concerning the meaning of legitimacy co-exist. Only through seeking answers to questions such as these can conflict and the exercise of power in society be understood. That is the essence of Weber's position in this respect. At one and the same time, therefore, sociological analysis is concerned to explain—through the use of type constructs—the connection between social action (with its emphasis on the actor's definition of the situation) and social structure (with the implication that social action cannot be understood in a totally free-

[1] *Theory*, op. cit., p. 124.

floating way). This is what is involved in Weber's contention that a sociological explanation has to be causally adequate and adequate at the level of meaning.

We will now seek to illustrate a social action approach by exploring what appears in everyday conversation as an undifferentiated term—'restrictive practices'. What we will attempt to show is that the approach takes seriously the explanation of actions offered by participants and situation, but that this does not exhaust the sociological task.

Restrictive practices as a form of traditional action

By a traditional form of action in the economic sphere we have in mind the type of employer or worker who simply wishes to earn sufficient money to enable him to carry on living in a way which for him is already established and acceptable. The implication is that changes in ways of working or in the character of market relationships (with customers and fellow competitors) are seen broadly as undesirable and given this kind of orientation are resented and resisted. We may observe that a certain solidarity between employer and employee based upon a traditionalistic orientation can occur in response to upsetting innovations. This is well illustrated in Hobsbawm and Rudé's study *Captain Swing*.[1] They note the markedly ambivalent attitudes of farmers and even magistrates to the destruction of threshing machines by farm labourers.

The Times, reporting on the first machine-breaking riots in East Kent (1830), observed: 'It is understood the farmers whose threshing machines have been broken do not intend to renew them. So far therefore the object of the riots will be answered. . . . Farms do not consider threshing machines of much advantage, seeing that they throw the labourers out of work and consequently upon the parish.'[2] Further, some farms destroyed their own machines even before the appearance of rioters. The traditionalistic spirit might expect to be reflected in popular opinion, and contemporary reports suggest widespread sympathy for machine breakers. In another paper, Hobsbawm cites examples of textile machine

---
[1] E. J. Hobsbawm and George Rudé, *Captain Swing*, Lawrence & Wishart, 1969.
[2] Quoted in Hobsbawm and Rudé, op. cit., p. 234.

workers in various parts of the country who were not brought to justice even though many masters knew well enough who had broken their frames; and he noted that in Rossendale merchants and woollen manufacturers had themselves passed resolutions against power looms some years before their employees smashed them. It is misleading to assume that employers necessarily supported technical progress over and against employees in periods of industrialization. Hobsbawm points out so far as England was concerned:

> The small shopkeepers or local master did not want an economy of limitless expansion, accumulation and technical revolution, the savage jungle pursuit which doomed the weak to bankruptcy and wage-earning status. His idea was the secular dream of all 'little men', which has found periodic expression in Leveller, Jeffersonian or Jacobin radicalism and small scale society of modest property owners and comfortably-off wage earners without great distinction of wealth or power; though doubtless in its quiet way getting wealthier and more comfortable all the time. It was an unrealizable ideal, never more so than in the most rapidly evolving of societies. Let us remember however that those to whom it appealed in early nineteenth-century Europe made up the majority of the population and outside such industries as cotton, of the employing class. But even the genuine capitalist entrepreneur could be in two minds about machines. The belief that he must inevitably favour technical progress as a matter of self-interest has no foundation. . . . Quite apart from the possibility of making more money without machines than with them (in sheltered markets, etc.), only rarely were new machines immediate and obvious paying propositions.[1]

However, employers who systematically organized themselves and their resources in the calculated pursuit of profit were another matter. Imbued with the capitalist spirit they stand out as anti-traditionalists. When business becomes the *raison d'être* of such a man's existence we have in type-form a rational orientation to a system of discrete individual ends. Such a man is no longer

[1] E. J. Hobsbawm, 'The Machine Breakers' in *Labouring Men: studies in the history of labour*, Weidenfeld and Nicolson, 1964, pp. 13–14.

satisfied with the traditional rate of profit or work. Weber has classically depicted the emergence of this type of man, and the character structure it produced, in *The Protestant Ethic and the Spirit of Capitalism*.[1] His anti-traditional outlook (whether rooted in a religious or utilitarian value system) necessarily shaped his attitudes towards those he employed. In pre-industrial England economic historians have drawn attention to the social habits of local communities which gave rise to voluntary under-employment and proliferation of saints' days (holidays), the instability of labour and irregular work patterns. In times of high wages the tendency to work short-time was common.[2] From such material the capitalist had to create a regular, disciplined, and industrious work force. Somehow one had to discourage or prevent people from worshipping at the shrine of St Monday (and possibly St Tuesday); that is—deal with the problem of systematic absenteeism. Equally there was the problem of getting an 'appropriate' amount of work from employees when they did turn up. A very early example of the difficulties is reflected in Thomas Crowley's *Law Book of the Crowley Iron Works* (1700). Order 103 makes reference to practices which would commonly now be described as restrictive and which were plainly anathema to Crowley:

> Some have pretended a sort of right to loyter, thinking by their readiness and ability to do sufficient in less time then others. Others have been so foolish to think bare attendance without being imployed in business is sufficient . . . Others so impudent as to glory in their villany and upbrade others for their dilligence. . . .[3]

Perhaps one of the most striking things which is noted in comments on traditional societies as compared with industrial societies is the attitude to time. Margaret Mead, writing about the Spanish Americans of New Mexico, USA, points out that they do not regulate their lives by the clock. It is possible to observe a round of seasonal activities set within an established framework

---

[1] M. Weber, *The Protestant Ethic and the Spirit of Capitalism*, Allen & Unwin, 1930.
[2] See, for example, D. C. Coleman, 'Labour in the English Economy of the Seventeenth Century', in *Economic History Review*, 2nd Series, VIII.
[3] Quoted in E. P. Thompson, *Time, Work, Discipline and Industrial Capitalism: Past and Present*, XXXVIII, December 1967, p. 81.

and linked to the calendar of sacred festivals and holy days.[1] As for work that was an accepted and inevitable part of everyday life, however, working hours are not specifically defined and traditionally work and rest are not seen in opposition to each other, but as part of the same process—work a little, rest a little. Further 'there is no moral corruption in being idle, or in staying away from one's job. A worker may stay away from wage-work, but may spend the day repairing a neighbour's door or helping to build a hen-house for nothing.'[2]

In the nature of the case one cannot assume that traditional forms of behaviour disappear overnight. Andreski, for example, points out that it would be surprising not to find a good deal of unpunctuality in recently industrialized societies. He observes: 'even today, wandering around the cities of Latin America, one notices that most of the clocks stand still or are inaccurate. This curious detail indicates either carelessness in tending mechanical objects or a lack of regard for time. Obviously, nowhere where the adage "time is money" is taken seriously would clocks be neglected.'[3] Yet industrial capitalism brings into being a form of work life in which time is 'spent', 'saved', 'added' and 'used'. These are all active words, far removed from notions of time 'passing'. In order to operate satisfactorily, industrial societies find it greatly convenient to operate according to 'standard time'. The discipline which is imposed on the working population is in part a time discipline. The attempt to impose such a discipline is inextricably bound up with attempts to impose appropriate standards of effort on the work force in the interests of 'efficiency'. The failure to do this to the employer's (and/or the government's) satisfaction, might be described as a failure to defeat traditionalism. The resulting behaviour is sometimes labelled 'restrictive'. Systematic absenteeism in mining and the irregular patterns of dockwork still found in industrial societies may have something of this character about them. Apart from casual employment patterns which de-casualization schemes seek to alter, work effort can be limited by such devices as 'spelling' or 'welting' in dockwork, which are described in the Devlin Report on the British Port Transport

[1] Margaret Mead, *Cultural Patterns and Technical Change*, Mentor, 1955.
[2] *ibid.*, p. 164.
[3] S. Andreski, *Parasitism and Subversion. The Case of Latin America*, Weidenfeld & Nicolson, 1966, p. 34.

Industry as highly organized forms of bad time-keeping.[1] The welting practice in Liverpool, described in the Report, related to the situation in which at any given time only half a gang was working. The others might be resting, smoking or having a tea-break. A similar practice in Glasgow was described as spelling. But it was also pointed out that 'in less organized forms bad time-keeping can become a practice in the sense that late starts, early finishes and prolonged tea-breaks can become so regular and so frequently condoned that attempts at enforcing punctuality would be resented perhaps to the point of unofficial action'.[2] A very striking example of the attempt to develop industrial discipline is found in post-revolutionary Russia. There, a mass campaign to encourage methodical work habits was undertaken by the Red Army. A Time League was formed to fight for 'the proper use and economy of time' which produced propaganda leaflets urging workers: 'Measure your time, control it! Do everything on time! exactly on the minute! Save time, make time count, work fast! Divide your time correctly, time for work and time for leisure.'[3] Such activity was derived from Lenin's conviction that, given the character of the earlier Tsarist regime together with a strong legacy of serfdom, the Russian was a bad worker.

The Effort Bargain

Writing of English industrial experience, E. P. Thompson comments:

> The first generation were taught by their masters the importance of time. The second generation found their short time committees in the ten-hour movement; the third generation struck for overtime or time and a half. They had captured the categories of their employers and learned to fight back within them. They had learned their lesson, that time is money, only too well.[4]

What is implied here is the growth of bargaining between workers and employers which is no longer described in terms of traditional

---

[1] Ministry of Labour, Final Report of the Committee of Inquiry under the Rt. Hon. Lord Devlin into certain matters concerning the Port Transport Industry. H.M.S.O., 1965. Cmd. 2734.
[2] *ibid.*, p. 19.
[3] Cited in R. Bendix, *Nation-Building and Citizenship*, Doubleday, 1964, p. 188.
[4] E. P. Thompson, *Time, Work-Discipline and Industrial Capitalism; Past and Present*, XXXVIII, 1967, p. 86.

versus instrumental rational orientations but in terms of competing rationalities.

A classical response of capitalist employers in dealing with the problem of work motivation and developing labour productivity was the promotion of individual piece-rate incentive schemes. This symbolized the employer's own view of the importance of the cash-nexus and his belief in the connection between the level of reward and degree of effort forthcoming from the worker. Any individualistic system of financial reward could also potentially undermine the traditional solidarity of the workers.

It is apparent, however, from the very earliest experiences that conflicts arose—complaints about speeding or rate cutting from workers and of loss of quality from employers and mutual recriminations over interpretations of the system and deductions made from pay for various reasons.[1]

Marx was one of the earliest to note that the piece-rate system resulted in constant friction between capitalist and labourer, and, in *Capital*, quotes from a contemporary study by Dunning on trade unions which by implication draws our attention to the relationship of behaviour which might be designated restrictive. Dunning's comment relates to the London engineering trade where a customary trick was 'the selecting of a man who possesses superior physical strength and quickness, as the principal of several workmen, and paying him an additional rate, by the quarter or otherwise, with the understanding that he is to exert himself to the utmost to induce the others who are only paid the ordinary wage to keep up with him... Without any comment this will go far to explain many of the complaints of stinting the action, superior skill and working power made by the employers against the men.'[2]

The example here is of an individual paid a (presumably secret) bonus to encourage the others. But even where all were on piece-rate incentive, problems were commonly encountered. Weber, for example, in his studies of German industry at the beginning of this century, noted that even workers who 'by virtue of their optimum efficiency can achieve and seek to achieve a wage considerably in excess of the normal measure... are usually compelled, directly or

[1] See, for example, Sidney Pollard, *The Genesis of Modern Management*, Penguin, 1968.
[2] Quoted in K. Marx, *Capital*, Allen & Unwin, 1938, p. 565.

indirectly, by the solidarity of their fellow workers, to "put the brake on", i.e. remain within the limits of average effort which will permit the others to "keep up", and which eliminates the danger, always in the workers' minds, of increases in earnings, due to particularly high output possibly causing the employer to reduce the piece rate'.[1] And in the American context around the same time Frederick Taylor asserted that ordinary piece-work systems promoted continual antagonism between employers and men. For him it is not the philosophy of economic incentives which is wrong but its application. Men may resist certain schemes not on traditional, but on economic grounds:

> Even the most stupid man, after receiving two or three piece-work cuts as a reward for his having worked harder resents this treatment and seeks a remedy for it in the future. Thus begins a war . . . between the workmen and the management. The latter endeavours by every means to induce the workmen to increase their output and the men judge the rapidity with which they work so as never to earn over a certain rate of wages, knowing that if they exceed this amount the piece-work price will surely be cut sooner or later. . . .[2]

Consequently Taylor was not surprised at the widespread existence of men holding back on work and 'marking time' or 'soldiering' as it was termed. This led to men being paid rates on what he regarded as a false basis, and he observed that such men would seek to obtain 'a soft snap', look after it, and earn as much as they felt they safely could without having the rate cut. Such behaviour was, from the men's point of view, economically rational. Taylor argued that well-managed day-work wage systems were preferable to piece-rate systems which brought such consequences. It would have been well perhaps if Taylor had left the matter there. There is a certain modernity about the position as is witnessed by Ford's decision to operate on time-rates rather than piece-rates. But Taylor argued that the problem arose because rate-fixing was done by guess-work and not based upon

---

[1] Max Weber, *Methodological Introduction for the survey of the selection and adaptation of workers in major industrial enterprises*, in J. E. T. Eldridge (ed.) Max Weber, *The Interpretation of Social Reality*, Michael Joseph, 1971, p. 133.
[2] F. Taylor, *Piece Rate System*, Routledge, 1919, p. 46.

accurate knowledge. Where the latter condition is fulfilled, conflict between employer and employee on this front is eliminated. Indeed 'one of the chief advantages . . . is that it promotes a most friendly feeling between the men and their employers, and so renders labour unions and strikes unnecessary'.[1] This simply ignores the fact that no matter how accurately one can time a job, one is still involved in a bargain over its price. Conflicts of interest are certainly not eliminated by 'scientific' rate-fixing (even if one permits the dubious assumption that there is one 'best way' of doing a job irrespective of the individuals performing it). For Taylor, however, any attempt to bargain over the impersonal 'objective' standard time, arrived at by accurate rate-fixing, was absurd. 'As reasonably might we insist on bargaining about the time and place of the rising and setting sun' was his response. Yet this cannot be dismissed as an historical oddity (though the view is rarely so starkly held today). Bell, for example, in his perceptive essay *Work and its Discontents*, pointed to the Aluminum Corporation of America's job evaluation programme designed to set wage differentials scientifically—with its diffident claim to be 'a mathematical tool for resolving day-to-day wage problems rationally and without dispute', and the widespread influence of methods-time-measurement schemes in the USA. From this standpoint, it becomes possible to define behaviour which opposes these 'objective' standards as irrational and, in principle, restrictive: 'For the engineer, the unwillingness of the worker to accept such a definition of a "fair day's work" only shows how deep-rooted is his irrational temper. The puzzled manager does not understand either why workers persistently restrict output and so limit their income. Yet the revolt against work is widespread and takes many forms.'[2]

Individual piece-rates are of course by no means a thing of the past. The National Board for Prices and Incomes' report on Payment by Results, in pointing this out, records the impression that group payment systems are not used to any large degree except where it is impracticable to distinguish the individual worker's contribution.[3] What is particularly interesting, Taylor's early

[1] F. Taylor, *Piece Rate System*, Routledge, 1919, p. 37.
[2] Daniel Bell, 'Work and its Discontents' in *The End of Ideology*, Free Press, 1965, p. 288.
[3] National Board for Prices and Incomes, Payment by Results Systems (HMSO, 1968), Cmd. 3627, p. 5.

critique of such schemes notwithstanding, is the report's observation that behind the technical differences between PBR schemes,

> ... most PBR systems in Britain whether of the traditional 'piece-price' or later 'time-base' variety, appear to be 'regressive' in the sense that effective price per unit declines as the worker's output rises. In the engineering industries for instance, where the base-rate on which the piece-work bonus is calculated has generally been much lower than the fall-back time-rate, and piece-workers' pay has often included a fixed national supplement, the relationship is nearly always less than proportional: gross earnings rise at a lower rate than output. Unit labour costs then declined with increased individual output, while the system also retains the advantage for the employer of protecting him in some degree against the higher unit costs brought about by a fall in output.[1]

What the report brings out, on the basis of its case studies, is that the shop floor (particularly where the labour force is predominantly male) was still described as a battlefield where the men and rate-fixers fight it out. Thus in one instance it was said that a good shop steward is one 'who spends at least fifty per cent of his time in the rate-fixing office arguing'.[2] This is in line with Lupton's earlier study of 'Jays Engineering' in *On the Shop Floor*, and also with the American study of Donald Roy.[3]

What might be termed the problem of traditionalism from the standpoint of the capitalist entrepreneur in relation to the labour force could be solved, so to speak, by putting a new heart within the worker: the heart of the economic individualist. This is the model non-restricting man. Broadly, what was wanted was a worker who internalized the bourgeois values of hard work, thrift, sobriety and ambition while remaining submissive to managerial control: accepting the employer's definition of the siutation so far as his expected output and wages were concerned. To find such

---

[1] National Board for Prices and Incomes, *Payment by Results Systems* (HMSO, 1968), Cmd. 3627, p. 5.
[2] *ibid.*, p. 19.
[3] T. Lupton, *On the Shop Floor* (Pergamon Press, Oxford, 1963). Donald Roy, 'Work Satisfaction and Social Reward in Quota Achievement: an Analysis of Piecework Incentive' in *A.S.R.*, XVIII, 1953; and 'Quota Restriction and Goldbricking in a Machine Shop' in *A.J.S.*, LVII, March 1952.

men is often to enter the world of the rate-buster. If he is in a minority among his workmates he is likely to be called a 'job-spoiler' by them. One well-known American study by Dalton of the industrial rate-buster (though based on a small sample) noted that most of the rate-busters came from urban lower-middle class families or from farms, whereas most of those who conformed came from urban working class backgrounds. He suggested that they were more ambitious than their fellows, placed great value on money, personal savings and aspired to attain a high socio-economic status.[1] Interestingly enough, the recent Luton study of affluent workers in the UK stresses their economic individualism, their willingness to 'follow the money'.[2] As with the Dalton study, it is suggested that such workers tended to originate from families of a higher socio-economic group than other workers. Another UK example of 'following the money' is Sykes' study of navvies,[3] although this was not accompanied by ideas of upward social mobility or indeed by savings—indeed the irregular and casual work patterns were a good example of 'traditionalistic' behaviour. But it is worth noting that in both cases the individualistic 'following the money' attitude represented a conditional attitude towards one's employer. For the navvy, it was an insult to be branded 'a Wimpey's man', 'a Costain's man' or whatever, and one asserted one's independence by 'jacking' a job in when one heard of a better paid site elsewhere. For the 'affluent worker', one remained with the firm for strictly instrumental reasons, so long as the conditions for obtaining high wages remained.

The tendency for women workers to be individualists who accept managerial definitions of the work situation has also been observed. The NBPI report on PBR for example comments:

> Our case studies have provided particularly striking evidence that management control of PBR is notably tighter where

---

[1] See M. Dalton, 'The Industrial "Rate-Buster": and Characterisation' in *Applied Anthropology*, VII, No. 1, 1948. See also, O. Collins, M. Dalton and D. Roy, 'Restriction of Output and Social Cleavage in Industry' in *Applied Anthropology*, Vol. V, No. 3, 1946.
[2] See John H. Goldthorpe et al., *The Affluent Worker: Industrial Attitudes and Behaviour*, Cambridge University Press, 1968.
[3] A. J. M. Sykes, 'Navvies: their Work Attitudes' in *Sociology*, III, No. 1, 1969. See also by the same author: 'Navvies: their Social Relations' in *Sociology*, III, No. 2, 1969.

women predominate in the labour force. It appears that women have not subjected such systems to the same degree of pressure as have men. On the whole, women appear to have accepted the result of work measurement as 'correct' and tend not to bargain over times or prices. In one firm the case study worker noted that 'when workers are being timed, far from their attempting to mislead by a carefully concealed slowing of work pace, many of the women are obviously too nervous for successful deceptions or, indeed, work faster than their normal pace as a matter of pride'.[1]

The individualistic stress on hard work is reflected here, as it was in Lupton's study of the 'Wye Garment Co.'.[2] But it is not the expectation or intention of upward social mobility that promotes such behaviour: rather it is a form of economic fatalism that one can do little to control the conditions of one's working life and the wages system in particular.

The industrial worker, however, can of course respond to the employer in group terms rather than on an individualistic basis. And that response does not have to be of the traditionalistic kind. What Thompson pointed to in British industrial history, as we have seen, is mirrored in the experiences of other industrializing and industrialized countries: men learn from their employer that time is money; they act and react in the light of that knowledge.

When this happens managerial definitions of the situation are not accorded absolute validity. More often than not they are contested. And it is the outcome of this contest that constitutes the effort bargain. There may in short be differing conceptions as to what constitutes 'a fair day's work'. What are termed restrictive practices may regulate the effort bargain. The point to stress is that, unlike behaviour adopted for traditionalistic reasons, these practices have an economic rationale. A circular justifying the Seaman's Ca'canny of 1896 observed for example: 'Employers of labour declare that labour and skill are "mere marketable commodities", the same as hats, shirts, or beef. . . . Then the possessors of such commodities are justified in selling their labour or skill in like manner as the hatter sells a hat or the butcher sells his beef.

---

[1] op. cit., p. 19.   [2] In Lupton, op. cit.

They give value for value. Pay a low price and you get an inferior article or a lesser quality.'[1]

In point of fact behaviour with a traditionalistic orientation may co-exist with behaviour which has a market orientation. This is brought out with particular clarity in Gouldner's study, *Wildcat Strike*.[2] Management saw a policy of rationalization as the only way to survive in a competitive gypsum mining industry following World War II. They abandoned the indulgency pattern which had hitherto characterized their labour relations and attempted to exercise tighter, more bureaucratic, control. The workers responded by restricting output, go-slows, higher absenteeism and eventually by striking.

The social disruption which the strike and allied worker responses symbolized is seen by Gouldner as a consequence of trying to 'free' the labour contract from the social forces which made it effective in the first place, namely the shared traditional beliefs and values in the community from which the workers derived their work expectations. The attempt by management to sharpen the terms of the labour contract with precise and often written statements about the duties and obligations of the worker served to highlight the conflict between these legalistic demands and the traditional social supports which had sanctioned worker obedience in the first place. For some of them it was the attempt to put the clock back to the old indulgency pattern days: to a situation in which the values of the community interpenetrated and were complementary to the values of the enterprise. For others it was an attempt to argue about the terms of the labour contract. Thus: 'If the traditionalists sought to return to a relationship governed by "trust", then the "market-men" desired a situation in which trust did not matter: they wanted their prerogative guarded by legal guarantee. If the traditionalists wanted to be able to return to the "fold" the "market man" wanted to be "taken into the business" '.[3]

The notion of giving 'value for value' comes out with particular clarity in Roy's study 'Quota Restriction and Goldbricking in a Machine Shop'.[4] The key question for a worker was whether he

[1] Quoted in E. H. Phelps Brown, *The Growth of British Industrial Relations*, Macmillan, 1959, p. 290.
[2] A. Gouldner, *Wilcat Strike*, Harper & Row, 1954.   [3] *ibid.*, p. 63.
[4] Roy, 'Quota Restriction', etc., op. cit.

could 'make out' on particular piece-rated jobs; that is reach a normal earnings level, in line with his expected take home pay. If he calculated that he could, he would often work very hard to do so. If he felt this was not possible, however, he would deliberately drop his effort to a minimum. This was an attempt to show the rate-fixer that the job was badly timed ('a stinker') and not worth the effort. It is worth noting, incidentally, that 'making out' on piece rates in the worker's mind had rather precise connotations in relation to the base-rate—one had to achieve at least 15 cents an hour above the base to justify the effort. The 'value for value' comment on a 'stinker' was: 'They're not going to get much work out of me for this pay', and: 'What's the use of pushing when it's hard even to make day rate?' This form of output restriction is described by Roy as goldbricking.

Restriction of effort could, however, operate on 'gravy jobs'. Here one is concerned with quota restrictions in order to preserve a loose rate. To achieve maximum earnings, it was argued, was something that would soon result in managerial corrective action: 'They'd retime the job so quick it would make your head swim! And when they'd retimed it they'd cut the price in half!'

Roy's analysis strongly supports the idea that the economic individualists are not the only economic men around! '. . . the operators in my shop make noises like economic men. Their talk indicated that they were canny calculators and that the dollar sign fluttered at the masthead of every machine. Their actions were not always consistent with their words. . . . But it could be precisely because they were alert to their economic interests—at least to their immediate economic interests—that the operators did not exceed their quotas. It might be inferred from their talk that they did not turn in excess earnings because they felt that to do so would result in piece-work price cuts: hence the consequences would be either reduced earnings from the same amount of effort expended or increased effort to maintain the take-home level.'[1]

The noises of these economic men were also heard by Roy and recorded in his paper 'Efficiency and "the Fix"'.[2] Roy describes an informal network in the lower reaches of the machine shop social structure (inspectors, set-up men, tool-crib men, stockmen

---

[1] Roy, 'Quota Restriction', etc., p. 430.
[2] Donald Roy, 'Efficiency and "the Fix": Informal Intergroup Relations in a Piecework Machine Shop' in *A.J. of S.*, LX, 1955.

and machine operators), who modified certain formally established shop routines, not to impede production but to facilitate it in order that the operatives might 'make-out'. Managerial attempts to eradicate the fix were, argued Roy, misguided in terms of their own goals. It is in this sense that he refers to the possibility of managerial inadequacies as a serious problem for labour. A similar argument is developed in *On the Shop Floor*, where the cross-booking system was condoned by immediate supervision because it was recognized and accepted as an attempt to 'make-out' to counteract delays and servicing difficulties which were properly blamed on inadequate managerial planning. In addition, we may note that whether the men on the shop floor behave like economic men or not, sometimes management will itself operate a quota restriction on production. Thus in a steel plant studied by the writer, which had instituted a progressive group bonus scheme with an accelerated bonus, once a certain tonnage was achieved, the men could only comment sarcastically upon such incentives because of the ceiling placed upon production by management planning. They referred to, and were not noticeably enthralled by, political exhortations to work harder for the sake of the balance of payments.

Now the central factor of the effort bargain is that it is subject to many and continuous processes. To this extent the employment contract is notoriously unstable. Different social groups may have different notions about what constitutes a fair day's work and an appropriate reward for a given level of effort. Writers such as Behrend[1] and Baldamus[2] have suggested that there are class and status differences. Certain conceptions of appropriate effort-reward levels may be generated and lead to management-worker conflict. Here we may draw attention to Baldamus' Wage-Effort paradigm (see chart). Looked at from the employee's point of view, wage parity exists where wage movements up or down are matched by proportionate effort changes (derived from his notions of a fair day's pay for a fair day's work).

The first five conditions of wage disparity are seen as advantageous to the worker, the second five as disadvantageous. Which conditions are found in particular cases will depend upon market, technological, and organizational factors, which both affect and are affected by the relative bargaining power of management and

[1] H. Behrend, 'A Fair Day's Work' in *S.J. of Pol. Econ.*, VIII, June 1961.
[2] W. Baldamus, *Efficiency and Effort*, Tavistock Publications, 1961.

BALDAMUS' WAGE-EFFORT PARADIGM*

		Wages	Effort
Wage disparity (positive)	(1)	+	—
,,	(2)	+	C
,,	(3)	C	—
,,	(4)	+ +	+
,,	(5)	—	—
Wage disparity (negative)	(6)	+	+ +
,,	(7)	— —	—
,,	(8)	C	+
,,	(9)	—	C
,,	(10)	—	+
Wage parity	(11)	+	+
,,	(12)	—	—

* Derived from *Efficiency and Effort*, pp. 118–19.

workers. Baldamus argues that it is a necessary condition for a successful administration to increase the level of average effort in a way that is not entirely offset by a corresponding rise in wages, whereas the worker at least wants to preserve wage parity. 'This conclusion, however, demands a wider interpretation of industrial strife than is customary. Not only strikes, but other kinds of instability, including absenteeism, excessive labour turnover, and restriction of output, have to be defined as symptoms of a fundamental discrepancy in the relation of effort to wages'.[1]

It is salutary to think on this matter in the light of the continuing discussion on the 'control of wages'. A situation which has got 'out of hand' from the standpoint of employers and the government turns out on examination to be one or other of the situations of positive wage disparity. So, for example, the NBPI report on PBR comments on the 'problem' of 'decaying' wage systems, which turn out on examination not to be simply a question of technical deficiencies (though it may be that as well) but points to a changing pattern of the distribution of rewards. Thus:

At one engineering firm we studied, the price of its main product has remained constant over the past few years while increased earnings have absorbed all the productivity gains derived from an expensive programme of technological change. At another firm a 'worn out' piece-work system produced an average annual

[1] W. Baldamus, *Efficiency and Effort*, p. 126.

increase in earnings of 6 per cent over the past five years; productivity was rising at the same rate and turnover increased 20 per cent but profit fell by 16 per cent. At a third firm average earnings must have risen by 75 per cent between 1963–67, much faster than the most optimistic estimate of a real increase in output, an increase which was in every case brought about by capital investment. In none of these instances was there any real attempt to control or reform the payment system. . . .[1]

It should be noted that one is not necessarily talking about a decline in plant productivity when one talks thus of a decaying wage system, but of a shift in the distributive pattern of the reward system, to the detriment of management and shareholders. Great play is made in the report on the significance of the 'earning' or 'improvement' curve whereby workers are able to improve performance on particular jobs with experience. It is argued that this should be taken into account in establishing work standards for new jobs and new workers or revising them for old jobs.[2] The deleterious effects on future improvement curves, as a result of this built-in disincentive, is not noted.

The attempt to re-assert managerial control on the wage front has, of course, been an integral part of productivity bargaining. But does it always deal with the problem of 'decay' as defined above? The NBPI report on Productivity Agreements, while it offers a stout defence of productivity (or, more broadly, efficiency) bargaining ends on a cautionary note: 'It would . . . be premature to conclude that productivity bargaining could not have an inflationary effect in the long term; time alone can show this. It is possible indeed that innovations in wage determination have their rise and fall; having for a time been beneficial they become subject to decay. . . .[3]

One may suggest that, in fact, political and industrial authorities tend to define the modes of worker behaviour which promote and maintain positive wage disparity as restrictive (and, of course, condemn 'weak managements' which permit it). Two important cases in the Baldamus Wage-Effort paradigm are (2) and (4). In (2)

---

[1] Payment by Results Systems, op. cit., p. 34.
[2] See p. 36 for example.
[3] National Board for Prices and Incomes. Productivity Agreements (HMSO, 1969). Cmd. 4136, p. 31.

THE SOCIAL ACTION PERSPECTIVE 61

wages increase and effort remains constant. This may occur as a result of technical change. This appears to have happened in one of the examples cited from the PBR report above. Baldamus also notes that 'the overall trend towards increased capital investment per unit of labour tends to diminish marginal wage disparity and thereby conditions the distributive process in favour of the employee'.[1] The third example in the PBR report cited above seems to fit here. This however might be offset by changes in the level of employment which, when reduced, could from the employees' point of view, give rise to an unfavourable relation of effort to earnings. In case (4) even though effort increases it is less than proportional to the wage increase. This, one may suggest, is the most likely illustration of a 'decayed' productivity bargain. By contrast, the 'successful' productivity bargain from the management standpoint is most likely to be located in case (6), where effort increases over-proportionately to the wage increase. This is because of the officially defined need to pass on gains to the consumer and the shareholders as well as to the workers themselves.

'Restrictive labour practices may be defined as "rules or customs which unduly hinder the efficient use of labour"'[2]. It will be useful at this point to refer to the Hawthorne investigation carried out at Western Electric, Chicago, in the late 1920s and 1930s.[3] In *Management and the Worker* a distinction is drawn between 'the logic of efficiency' and 'the logic of sentiments'. The logic of efficiency refers to the articulated set of beliefs about how one may organize to optimum advantage the firm as an economic unit and co-ordinate the factors of production accordingly. The logic of sentiments refers to the values residing in the inter-human relations of the different groups within the organization. Examples of what is meant here are the arguments employees give which centre upon the 'right of work', 'seniority', 'fairness', 'the living wage'.[4] This is a useful distinction because it draws attention to the fact that even though the logic of sentiments may impede the operation of the logic of efficiency what we are in principle discussing is not *rational* (i.e. efficiency-orientated) versus *irrational*

---

[1] Baldamus, op. cit., p. 110.
[2] Royal Commission on Trade Unions and Employers Associations (HMSO, 1968), Cmd. 3623 (Donovan Report), p. 77.
[3] F. J. Roethlisberger and W. J. Dickson, *Management and the Worker*, John Wiley & Sons Inc., New York, 1964.
[4] *ibid.*, p. 564.

behaviour, but rather *co-existing forms of rational behaviour operating from different premises*. The application of the distinction is also useful, because while arguing that management groups are more likely to be identified (formally) with the efficiency goals of the organization, nevertheless 'all groups within . . . participate in these different logics'.[1] Very good examples of the logic of sentiments pursued by managers, and modifying their efficiency orientation, are to be found in Burns and Stalker, *The Management Innovation*[2] and Dalton, *Men Who Manage*.[3] Both studies, for example, stress the way in which information may be withheld as a means of defending one's status, or undermining another person's authority, together with tactics designed to maintain or enhance one's control over resources. Dalton tends to formulate management in terms of intrigue. But it need not be hypocritical, for, as Burns and Stalker point out:

> The fact that interests of the firm may best be served, according to the Sales Director, by building up the sales force, or, according to the Works Manager, by putting in plant or machinery, or, according to the Research Director, by taking on more development projects, is not, they feel, material, and to impute that it is in itself discreditable is unfair.[4]

Each specialist in fact tends to see the total organization from the aspect of the logic of his own speciality.

When one talks of rules or customs hindering the efficient use of labour (including managerial labour) one has to come to terms with the relative nature of the efficiency concept. This, we have already seen, depends very much upon the social standpoint of the observer and his occupational interests. In what is sometimes quoted as a classic case of output restriction, the Bank Wiring Observation Room study, the research workers take trouble to point out that, while the men's 'bogey' was below management standards, compared to the output of men in other concerns doing the same kind of work, it was high:

[1] F. J. Roethlisberger and W. J. Dickson, *Management and the Worker*, John Wiley & Sons Inc., New York, 1964, p. 565.
[2] T. Burns and G. Stalker, *The Management Innovation*, Tavistock Publications, 1961.
[3] M. Dalton, *Men Who Manage*, John Wiley & Sons Inc., New York, 1969.
[4] Burns and Stalker, op. cit., p. 145.

The average output per man in outside concerns was about 4,000 connections per day as compared with 6,000 for this group. The department officials were proud of their accomplishments and some of them commented that if the men consistently turned out more than 6,000 connections a day they would 'wear their fingers out'. Any outsider watching the men work . . . would have concurred in this statement. The speeds attained by some of the men were in fact astonishing. It is well to remember, therefore, that only with reference to an abstract logic of efficiency could the phenomenon described be called 'restrictive'.[1]

Now the existence of different efficiency levels with reference to a particular activity has to be understood situationally, that is—with reference to the striking of particular effort bargains.

It is worth recalling that, in his note of reservation to the Donovan Report, Shonfield argues that one should make efficiency comparisons of this sort, in order to lift the lower groups up to the standard of the highest. But it surely has to be recognized that such an endeavour has to be seen in a bargaining context, and, indeed, the more general socio-economic location of the plant and its workers. One cannot mechanically apply the criterion that because efficiency methods are already in operation elsewhere, which 'can be shown not to result in unusual strain or discomfort to the workers involved',[2] they must automatically be adopted. Shonfield himself points out that one would not necessarily adopt efficiency methods used by overseas competitors, because they might conflict with some accepted standards of social behaviour in our own country. This is so, and is a reminder that considerations of efficiency do not operate in a social vacuum. Shonfield does not add that conflicting notions of acceptable standards of behaviour exist *within* a society. Thus, even if one assumes that management and labour in a company are efficiency-orientated, it does not follow that they will agree with the legitimacy of the means adopted by management to achieve their ends. And where they do agree there is still a bargain to be struck. The fact that productivity bargains have often involved 'buying out' what have come to be regarded as restrictive practices, illustrates the point adequately.

The allied issue of technical change and workers' attitudes

[1] Roethlisberger and Dickson, op. cit., p. 537.
[2] Donovan, op. cit., p. 296.

towards it may be confronted in socio-political terms whilst recognizing the importance of the firm in providing a climate in which there is a tendency to resist or adjust effectively to change. Many of the difficulties are external to the firm itself so far as the worker is concerned. Thus the problem of restrictive practices as resistance to change may be redefined. This was clearly argued in the Donovan Report.

> The fears associated with change are reasonable fears. For change may face people—and their families too—with urgent practical problems. Technological change and the expansion of new industries and the decay of old, and the need for mobility of labour mean that men are obliged to change jobs, and perhaps to move to completely new areas to live and work. They will find the skills they possess are no longer in demand. They may face a severe cut in pay and so on. These are substantial problems and so long as they go unsolved the introduction of change itself will go unsolved.[1]

(4) Social Action and Reference Group Analysis

According to Merton, 'reference group theory aims to systematize the determinants and consequences of those processes of evaluation and self-appraisal in which the individual takes the values or standards of other individuals or groups as a comparative frame of reference'.[2] In Britain the debate about the embourgeoisement of the working class has to some extent been clarified by the application of reference group analysis, notably in the writings of Lockwood, Goldthorpe and Runciman.[3] In 'Affluence and the British

---

[1] Donovan, op. cit., p. 77.
[2] R. Merton, *Social Theory and Social Structure*, Free Press, 1957, p. 234.
[3] John H. Goldthorpe and D. Lockwood, 'Affluence and the British Class Structure' in *Sociological Review*, XI, 1963; John H. Goldthorpe *et al.*, 'The Affluent Worker and the Thesis of Embourgeoisement: some preliminary

Class Structure', Goldthorpe and Lockwood suggest that one might try to separate manual workers according to whether they exhibit working class or middle class normative identifications and then further distinguish between those whose reference group is their membership group, so far as actual social relations are concerned. The full sequence of embourgeoisement is portrayed as having four stages:

Traditional Worker ———⟶ Privatized Worker ———⟶
Socially Aspiring Worker———⟶ Assimilated Worker.

This is seen as the process of 'assimilation through aspiration' into the middle class. They argue that what has been loosely thought of as assimilation ('we're all middle class now') is much more properly interpreted as privatization or, at the most, social aspiration. This argument based on contemporary community studies has since been supported by their own empirical work at Luton which pointed 'fairly clearly to a considerable degree of status segregation'[3] on a class basis.

At this point one may introduce a distinction between Merton's advocacy of reference group theory and Lockwood and Goldthorpe's utilization of it. Merton overtly links reference group analysis to functional theory, suggesting that positive orientation towards the norms of the non-membership group may have functional or dysfunctional consequences for the individual depending upon whether the social structure is of an 'open' or 'closed' character.[2] However, Lockwood and Goldthorpe link their reference group analysis to a social action perspective. Thus, while they depict a mode of assimilation through aspiration to middle class norms, they recognize that the model itself is only a first step to clarity. They explicitly discuss the phenomenon of normative convergence. The 'radical individualism' of the middle class is seen as being replaced, to some degree, by an 'instrumental collectivism', which this class has come to accept as a more effective way of maintaining or improving the economic position of one's family. But working class behaviour was, perhaps, moving to this position from the traditional working class community-anchored

research findings' in *Sociology*, I, 1967. W. G. Runciman, *Relative Deprivation and Social Justice*, Routledge and Kegan Paul, 1966.
[1] Goldthorpe *et al.*, op. cit., p. 22.   [2] Merton, op. cit., p. 266.

'solidaristic collectivism', since among manual workers the economic and social advancement of the individual nuclear family has come to be accepted in many cases as a primary goal. One ought perhaps to recognize, so far as manual workers are concerned, that, in so far as one can discern a movement from 'solidaristic collectivism' to 'instrumental collectivism' the tendency might not be irreversible. Thus Luton is a town which has experienced a high level of inward migration of labour, mainly because of job opportunities in the car industry. It remains possible that a greater degree of residential stability will lead to a greater degree of communal sociability. Hence the privatized worker of today *may* be the traditional worker of tomorrow.

There is however a further refinement in the reference group analysis, discussed above, which now needs to be introduced. One assumption of the Goldthorpe and Lockwood paper is that one can, at a very general level, distinguish between working-class perspectives and middle-class perspectives, in terms of beliefs and values about society and one's place in it. Basically one had a working-class perspective, or a middle-class perspective, or a 'new' perspective (the position of normative convergence). In the original model the 'traditional worker', by definition, viewed society from *the* working-class perspective. However, Lockwood himself has subsequently drawn attention to the over-simplification implied here, suggesting that one should properly draw a clear distinction between two types of traditional worker—the 'proletarian' and the 'deferential'.[1] It was the proletarian traditional worker who figured in the first analysis. He is located most clearly in 'occupational communities', described by Lockwood as a work-dominated collectivity: 'the isolated and endogamous nature of the community, its predominantly one-class population and low rates of geographical and social mobility all tend to make it an inward-looking society and to accentuate the sense of cohesion that springs from shared work experiences'.[2] Mining is the clearest example of such a community and the much quoted study by Dennis *et al.*[3] has provided us with a very fruitful case study.

The deferential traditional worker, by contrast, is most likely to be found in work situations where there are hindrances of one sort

---

[1] D. Lockwood, 'Sources of Variation in Working-Class Images of Society' in *Sociological Review*, XIV, 1966.
[2] *ibid.*, p. 251.   [3] Dennis *et al.*, op. cit.

or another to the formation of strong ties with workers in a similar market situation to his own, as for example in service occupations, small family businesses and agricultural employment. The essence of his work situation is typically one in which 'the relationship between employer and worker is personal and particularistic'.[1] Such work-place relations are notably reinforced, Lockwood argues, in 'small relatively isolated and economically autonomous communities, particularly those with well differentiated occupational structures and stable populations.'[2] The point then is that one cannot talk of the blue collar worker's traditional image of society and contrast it with the 'new' worker's image, or the middle-class image but rather must locate different traditional images to particular structural considerations. And it is important to note that Lockwood's views are not pure speculation since they are based upon the idiographic evidence of existing local community studies. If then the proletarian traditional worker advocates collective action as a response to the power of 'them', the deferential traditional worker quietly accepts social inequality as a fact of life and the allocation of rewards (honour, prestige and income) to the existing status hierarchy as legitimate. As usual, of course, one enters the caveat that the structural factors promoting one or another of these images of society are not always found in their pure form.

Much of the discussion surrounding the embourgeoisement thesis derived from statements to the effect that affluent workers were becoming middle class in outlook as a result of their prosperity and would, therefore, increasingly tend to vote Conservative. Would not, therefore, Tory governments continue to be elected? This was the fashionable question for political commentators to pose in the late 1950s: the 'never-had-it-so-good' era enjoyed by the individualistic 'I'm all right Jack'. Now the Lockwood distinction between the two types of traditional worker is a reminder of one long-standing base of working class Conservatism: the deferential voter. However, Runciman has shown that even the deferential voter is not altogether a homogeneous category.[3] It might be thought, for example, that deferential working class Tories were those who also rated themselves as working class, whilst 'socially aspiring' working class Tories identified themselves as middle class.

[1] Lockwood, op. cit., p. 253. [2] *ibid.*, p. 253.
[3] Runciman, *Relative Deprivation and Social Justice*, op. cit.

Runciman shows that in fact 'those who describe themselves as "middle class" are more likely, not less, to give a reason suggestive of "deference" than are those who describe themselves as "working class"'.[1] However, manual workers who rate themselves as working class and vote Conservative tend to do so 'out of inherited habits and primary group loyalty'.[2] It is possible that the deferential acceptance of the Tories' fitness to govern among manual workers who rate themselves middle class may be of a more conditional kind. This is the calculative voter who figures that one party will meet his expectations of increasing prosperity more effectively than another. Support for this possibility appears in Runciman's earlier paper where, in discussing manual workers who rated themselves middle class, he locates the majority of them in the 'privatization' category of the Lockwood and Goldthorpe schema, rather than the 'aspiring' or 'assimilated' positions.[3]

Now it is the privatized worker whom Lockwood, in the article cited above, contrasts with both the proletarian and the deferential traditional manual worker. He holds neither a 'power model' nor a 'status model' of society but a 'pecuniary model': 'the logic of a purely pecuniary model of society leads to neither class consciousness nor status consciousness but to commodity consciousness ... at work he is wage oriented and in the community consumption oriented.'[4]

The voting behaviour of this particular kind of blue collar man, because it is calculative, is not to be pinned once and for all to any one party. The political implications are then quite different from a conception of working-class affluence (where it actually exists) resulting in the 'socially aspiring Conservative voter'—the manual worker who votes Conservative because of the higher status which he feels his action symbolizes. (Interestingly enough, Goldthorpe *et al.* found no evidence at all in Luton of the existence of the socially aspiring Conservative.)[5] It may be suggested that British electoral experience in the 1960s was consistent with the preceding argument.

One final point may be made in this section. Cannon[6] in his study

---

[1] Runciman, *Relative Deprivation and Social Justice*, op cit., p. 181.
[2] *ibid.*, p. 181.
[3] Runciman, *Embourgeoisement, Self-Rated Class and Party Preference*, op. cit.
[4] Lockwood, op. cit., p. 261.   [5] Goldthorpe *et al.*, op. cit., p. 27.
[6] I. C. Cannon, 'Ideology and Occupational Community: a study of Compositors' in *Sociology*, I, No. 2, 1967.

of compositors, has indicated that, while the embourgeoisement process has partially to be understood as a process of working-class people changing their norms, standards and aspirations, the normative concept itself is not unitary but involves a number of dimensions. He shows, for example, that the compositor 'not only accepts certain norms such as home ownership, but also has other values associated with the middle class, including educational and occupational aspirations for his children: yet . . . a high degree of working-class affiliation and labour party support is found amongst compositors'.[1] What has to be recognized here is that the occupational community serves as an important membership and reference group in explaining the continued working-class identification. To describe these workers as privatized would, therefore, be misleading. This suggests that the occupation itself needs to be treated as an intervening variable in assessing the extent to which one can talk about normative convergence between the working class and the middle class.

[1] I. C. Cannon, 'Ideology and Occupational Community: a study of Compositors' in *Sociology*, I, No. 2, 1967, pp. 182–3.

Part Two

*Industrial Society:
Anomie—and a New Integration?*

(1) *Durkheim: Anomie and Economic Life*

The concept of anomie is not alone in having a chequered and sometimes confusing history in sociological thought. It is our purpose in this chapter to trace out some of the varieties and, indeed, ambiguities of its history, and, in particular, to discuss the way in which it has served to interpret economic life.

If we begin almost inevitably with Durkheim it is not enough to remind ourselves of certain familiar points, but to draw attention to one or two relatively neglected ones—not least because one suspects that there are vulgar Durkheimians as well as vulgar Marxists. Durkheim was of course preoccupied with explaining how it is that societies and social groups can and do hold together at all as structures, and with delineating and accounting for the basis upon which individuals are prepared to act in concert with others. These closely related problems of social cohesion and social solidarity were first confronted by Durkheim in *The Division of Labour and Society*.[1]

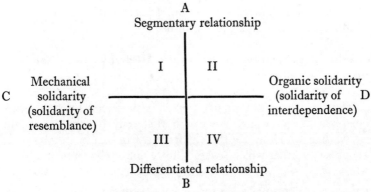

[1] E. Durkheim, *The Division of Labour in Society*, Free Press, 1964.

Part of Durkheim's thesis may be noted with reference to the diagram on page 73.

To the question how are societies and groups held together, Durkheim points to the alternative of structures dominated by the segmental principle and those characterized by differentiated relationships. In commenting on the segmentary principle Durkheim uses a speculative evolutionary approach by postulating first (as Spencer had done) the *horde* (the most elementary aggregate in which the 'parts' that go to make up the unit are indistinguishable from each other) as 'the veritable social protoplasm, the germ whence would arise all social types';[1] then the *clan*, which implies a linking of hordes to form a more extensive group; and eventually, in this respect, the *segmental society with a clan base*.

> We say of these societies that they are segmental. In order to indicate their formation by the repetition of like aggregates in them, analogous to the rings of an earth worm. . . .[2]

Corresponding to this segmental structure is a mechanical form of solidarity (Position 1 in the diagram). It implies that the individual is integrated into society at the expense of any individuality he may have been able to, or wanted to, express.

> Solidarity which comes from likenesses is at its maximum when the collective conscience completely envelops our whole conscience and coincides with all points in it. But at that moment, our individuality is nil. It can be borne only if the community takes a small toll of us.[3]

Why does Durkheim term the solidarity of resemblance mechanical solidarity?

> The term does not signify that it is produced by mechanical and artificial means. We call it that only by analogy to the cohesion which unites the elements by an inanimate object, as opposed to that which makes a unity out of elements of a living

---

[1] E. Durkheim, *The Division of Labour in Society*, Free Press, 1964, p. 174.
[2] *ibid.*, p. 175.   [3] *ibid.*, p. 130.

object. ... The individual conscience, considered in this light, is simply dependent upon the collective type and follows all of its movement, as the possessed object follows those of its owner.[1]

The individual is then seen as being bound to society without the operation of any intermediary agency.

A social structure with differentiated relationships as its basis is contrasted with a segmental system, and this is closely linked with Durkheim's notion of organic solidarity—the solidarity which he holds to be produced by the division of labour (and represented by position 4 in the diagram). The structure consists of 'a system of different organs each of which has a special role, and which are themselves formed of differentiated parts'.[2]

Mechanical solidarity

is possible only in so far as the individual is absorbed into the collective personality; the second is possible only if each one is a sphere of action which is peculiar to him; that is a personality. It is necessary then, that the collective conscience lean upon a part of the individual conscience in order that special functions may be established there, functions which it cannot regulate. The more this region is extended, the stronger is the cohesion which results from this solidarity.[3]

Durkheim chose to speak of organic solidarity in depicting this situation because he says that it is analogous to a higher animal where 'each organ in effect has its special physiognomy, its autonomy. And moreover the unit of the organism is as great as the individuation of the parts is more marked.'[4]

Already we can see that Durkheim has attempted (albeit in perhaps a fatally abstract form) a neat solution to the problem of integration. Position 4 in the diagram represents a coincidence of social integration (the individual is linked to society in a way which permits the growth of his individuality) and system integration ('society becomes more capable of collective movement, at the same time that each of its elements has more freedom of movement'.)[5] It is this situation which Durkheim regards as

---

[1] E. Durkheim, *The Division of Labour in Society*, Free Press, 1964, p. 30.
[2] ibid., p. 181.   [3] ibid., p. 131.   [4] ibid., p. 131.
[5] ibid., p. 131.

'normal', 'healthy' and 'desirable' in modern society: the adjectives are interchangeable and one should note the potentially misleading sense of the term normal here. In presenting a law-like statement about social development moving from position 1 to position 4 (using a very dubious index of repressive law to indicate mechanical solidarity, and restitutive law to indicate organic solidarity) it becomes evident not only that Durkheim is unable to find a precise historical example of mechanical/segmental society, but also that the organic/differentiated society 'is nowhere observable in its absolute purity'.[1] They are 'ideal-type' (hypothetical) formulations representing two extremes of social development. Position 1 represents the 'best fit' for a society based on mechanical solidarity and position 4 a 'best fit' for societies based on organic solidarity. This means that actual societies may sometimes correspond more closely to positions 2 and 3 despite the fact that in Durkheim's terms they embody structural incompatibilities and antagonisms. So, for example, England, which in the nineteenth century revealed a high degree of differentiation in the economic sphere, was also said by Durkheim to be composed of a large number of segmental organized local communities.

However, for our purposes, it is enough to point out that when Durkheim writes of abnormal forms of division of labour it is in the most general sense a recognition that the desirable congruence between system integration and the social integration of the individual is not always present in modern industrial societies. Organic solidarity, that is to say, is only imperfectly realized. This prompts Durkheim to reflect upon the role of the State as a regulating agency and the role of occupational organizations as an integrating factor. The two themes are closely related:

> ... the day will come when our whole social and political organization will have a base exclusively, or almost exclusively, occupational ... this occupational organization is not today everything that it ought to be ... abnormal causes have prevented it from attaining the degree of development which our social order now demands.[2]

Durkheim goes on to distinguish further between various categories of abnormality, one of which is the anomic division of

[1] E. Durkheim, *The Division of Labour in Society*, Free Press, 1964, p. 190.
[2] *ibid.*, p. 190.

labour. In structural terms Durkheim is here referring to the failure of a traditional society to adjust adequately to the conditions of modern economic life, and which commercial and industrial crises—conflicts between capital and labour (and within the ranks of labour) in certain of their aspects—may be regarded as expressing. The increasing physical separation of the producer from the consumer is regarded as of particular significance.

> The producer can no longer embrace the market in a glance, or even in thought. He can no longer see its limits, since it is, so to speak, limitless. Accordingly production becomes unbridled and unregulated. It can only trust to chance and in the course of these gropings it is inevitable that proportions will be abused, as much in one direction as in another. From this come the crises which periodically disturb economic functions'.[1]

The factory system with its emphasis on machines and manufacturing is closely linked to changes in market relations—and is seen by Durkheim to alter the relations between employers and employees and to separate the worker from his family (as opposed, say, to agricultural or domestic systems of production). It is in the context of this form of abnormality that the effects of the division of labour on the worker are spelled out:

> The division of labour has often been accused of degrading the individual by making him a machine. And truly, if he does not know whither the operations he performs are tending, if he relates them to no end, he can only continue to work through routine. Every day he repeats the same movements with monotonous regularity, but without being interested in them or without understanding them. . . . He is no longer anything but an inert piece of machinery, only an external force set going which always moves in the same direction and in the same way. Surely, no matter how one may represent the moral ideal, one cannot remain indifferent to such a debasement of human nature.[2]

---

[1] E. Durkheim, *The Division of Labour in Society*, Free Press, 1964, p. 370.
[2] ibid., p. 371.

The stress here is on the subjective sense of the meaninglessness of work arising from the fact that the new form of organic solidarity has not yet been fully developed to meet changed economic conditions. By arguing in this way, Durkheim is making clear that it is not the division of labour as such that is the cause of social ills.

It is, however, of great importance to recall that Durkheim refers also to the forced division of labour as another abnormal form. While conflicts of interest between employers and employees could be partly interpreted as a product of an anomic situation, which in time could be reconciled into a working equilibrium, that was not all. They were also partly a product of *'the still very great inequality of the external conditions of the struggle'*.[1] What is meant by this statement? The division of labour only produces the solidarity that it is capable of when each individual does the task to which he is fitted. If there is a mis-match between the aptitudes of individuals and their actual activities then 'only an imperfect and troubled solidarity is possible'.[2] The crux of Durkheim's argument is in the following passage.

> ... we may say that the division of labour produces solidarity only if it is spontaneous. . . . But by spontaneity we must understand not simply the absence of all express violence, but also of everything that can even indirectly shackle the free unfolding of the social force that each carries in himself. It supposes, not only that individuals are not regulated to determinate functions by force, but also that no obstacle, of whatever nature, prevents them from occupying the place in the social framework which is compatible with their faculties. In short, labour is divided spontaneously only if society is constituted in such a way that social inequalities exactly express natural inequalities. But, for that, it is necessary and sufficient that the latter be neither enhanced nor lowered by some external cause. Perfect spontaneity is, then, only the consequence of another form of this other fact—absolute equality in the external conditions of conflict.[3]

Durkheim recognizes that a society characterized by perfect spontaneity does not exist in any complete sense, and draws

---

[1] E. Durkheim, *The Division of Labour in Society*, Free Press, 1964, p. 370.
[2] *ibid.*, p. 376. [3] *ibid.*, p. 377.

attention to the hereditary transmission of wealth as a very important factor promoting inequality in the external conditions of conflict. This, he sees, has applications not only for individuals but also groups: the point which he develops in his discussion of contractual relations. A formally free contract between parties is not sufficient to provide the basis of organic solidarity. Durkheim notes the possibility of exploitation if a class has to take any price for its services in order to live, and argues in particular that 'there cannot be rich and poor at birth without there being unjust contracts'.[1] Durkheim's discussion of the forced division of labour enables him to draw a distinction between regulation based upon constraint—a feature of the forced division of labour—and regulation as an aspect of 'true' organic solidarity. If the first is the enemy of freedom, the second is the basis of it—for only through such regulation can external conditions promoting inequality be diminished.

So far, then, the matter comes to this: the spontaneous character of the division of labour tends to be blemished in conditions of anomie represented by an absence of social regulation or, minimally, an inadequate regulation of social life; and/or by a situation

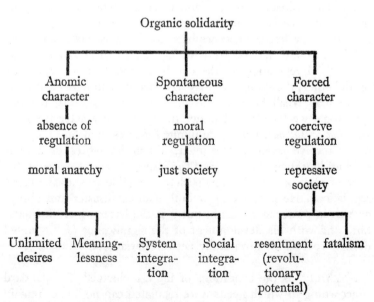

---

[1] E. Durkheim, *The Division of Labour in Society*, Free Press, 1964, p. 384.

in which force keeps a social order together imposed externally, so to speak, upon the division of labour.

What emerges from this is that any task of social reconstruction has to recognize these two analytically distinct sources of social unrest. If anomic situations breed unbridled conflict, coercive situations attempt to abolish conflict. Both inject a precarious element into organic societies. Durkheim maintains that 'it is neither necessary nor even possible for social life to be without conflicts. The role of solidarity is not to suppress competition but to moderate it'.[1]

However, it is also relevant to recognize that from the moral anarchy of the anomic division of labour, two separate consequences are implied: one is that the absence of regulation can lead to continually unspecified desires, and the other is that for the individual specialist worker, work itself can lose its meaning. Likewise, in the case of the forced division of labour, there are two possible consequences: one is a resentment of exploitation and an attempt to meet force with force, the other is a fatalistic acceptance of domination. One can readily see how, in a society with the anomic and forced division of labour co-existing, the unlimited desires/resentment and meaninglessness/fatalism pairs might reinforce each other, which helps to explain why anomie and alienation have sometimes been confused as social categories. While they are derived from different structural arrangements (absence of regulation on the one hand and coercive regulation on the other) the symptoms may have certain common elements.

We may recall here that in his later study *Suicide*, the two symptoms we have noted under the anomic character from the *Division of Labour*, namely 'unlimited desires' and 'meaninglessness', receive separate treatment. When he writes of anomic suicide, Durkheim has in mind the impact of social crises on the lives of men, at which point society is unable to exercise moral regulations. He is thinking of both sudden disaster and abrupt growth in power and wealth, but it is the latter which fascinates him, and with his development of the significance of 'unlimited desires' reveals his profoundly anti-utilitarian convictions,

> ... truly, as the conditions of life are changed, the standard according to which needs were regulated can no longer remain

[1] E. Durkheim, *The Division of Labour in Society*, Free Press, 1964, p. 365.

the same. . . . The scale is upset; but a new scale cannot be immediately improvised . . . The limits are unknown between the possible and the impossible, what is just and what is unjust, legitimate claims and hopes and those which are immoderate. Consequently there is no restraint upon aspirations. . . . Some particular class especially favoured by the crises is no longer resigned to its former lot, and, on the other hand, the example of its great good fortune arouses all sorts of jealousy below and about it. Appetites, not being controlled by a public opinion become disoriented, no longer recognize the limits proper to them. . . . With increased prosperity desires increase. At the very moment when traditional rules have lost their authority, the richer prize offered these appetites stimulates them and makes them more exigent and impatient of control. The state of de-regulation or anomy is thus further heightened by passions being less disciplined, precisely when they need more disciplining.[1]

And it is the sphere of economic life which Durkheim, writing about nineteenth-century industrial Europe, diagnoses as being in a state of acute anomie:

There the state of crises and anomy is constant and, so to speak, normal. From top to bottom of the ladder, greed is aroused without knowing where to find ultimate footholds. Nothing can calm it, since its goal is far beyond all it can attain. Reality seems valueless by comparison with the dreams of fevered imaginations; reality is therefore abandoned, but so too is possibility abandoned when it in turn becomes reality.[2]

However, the category of meaninglessness is linked by Durkheim not to anomic but to egoistic suicide. This is treated by Durkheim as a product of excessive individualism. Such a person is withdrawing from group ties and more and more basing his conduct on his own private interests. But as one withdraws—then, in a sense, one withdraws from the moral rules, laws and dogmas of society. As one asks, for what purpose do they exist, one also asks, for what purpose do I exist? The individual, because he is not integrated with society, comes to lose his social identity:

[1] E. Durkheim, *Suicide*, Routledge & Kegan Paul, 1952, pp. 252–3.
[2] *ibid.*, p. 256.

Social man necessarily presupposes a society which he expresses and serves. If this dissolves, if we no longer feel it in existence in action about and above us, whatever is social in us is deprived of all objective foundation. All that remains is an artificial combination of illusory images, a phantasmagoria vanishing at the least reflection: that is nothing which can be a goal for our action. Yet this social man is the essence of civilized man; he is the masterpiece of existence. Thus we are bereft of reasons for existence. . . .[1]

So it is that 'absence of regulation' can imply the continual searching for fresh goals and aspirations which are ever to remain unsatisfied; or the withdrawal of the individual from the pursuit of social goals. In *The Division of Labour* this sense of purposelessness was subsumed as part of the anomic division of labour. In *Suicide* the distinction is very clear cut. Whilst allowing that anomic and egoistic suicide have certain affinities, 'since both spring from society's insufficient presence in individuals', nevertheless 'one may live in an anomic state without being egoistic and vice versa'.[2] There is, however, one difference in Durkheim's presentation of the concept of purposelessness as between *The Division of Labour* and *Suicide*. In *Suicide* the purposeless individual comes to doubt the value of life through an intellectual process—he detaches himself progressively from socially constructed goals till he has none left to which he can commit himself. In *The Division of Labour*, the purposeless individual is the labourer who has no goals which are truly his own—he is the slave of routine—and the suggestion is that in such a situation he is scarcely capable of thought. Rather he is so habituated in his actions that he is more like a piece of machinery than a human being.

But there remains another and perhaps more precise parallel in Durkheim's treatment of suicide as an expression of social disorganization and the abnormal division of labour. Durkheim contrasts anomic suicide with fatalistic suicide:

> . . . it is the suicide deriving from excessive regulation, that of persons with futures piteously blocked and passions violently choked by oppressive disciplines.[3]

---

[1] E. Durkheim, *Suicide*, Routledge & Kegan Paul, 1952, p. 213.
[2] *ibid.*, p. 258.   [3] *ibid.*, p. 276.

He does not see this category as being of much contemporary significance. At the same time he sees fatalistic suicide occurring under excessive physical or moral despotism which certainly has strong affinities with his earlier comments in the *Division of Labour* on the forced division of labour. If one did not escape from or was able to overthrow the exploiters under the forced division of labour, the fatalistic response, whilst not necessarily suicidal is similar in kind.

What is to be done? The task of social reconstruction

How can we hope to have moral men in a moral society, particularly a modern industrial society? This involves a consideration of what part the economic order should play in relation to the rest of social life, and, so far as the individual is concerned, it implies for Durkheim that social justice is the prerequisite for the integration of the individual into society. What is needed is a policy which would draw the egoist out of his social isolation, curb the appetites of the unsatisfied strivers, and remove the conditions which promote and sustain unequal competition. Several things emerge as Durkheim attempts in his work to consider the implications of such reconstruction.

(1) First, a policy which operates solely to remove social inequalities, while desirable in the interests of social justice, is only a partial solution. Subsidies, relief and welfare policies to diminish social inequalities are desirable but in themselves inadequate, and, taken on their own, may serve to compound the problem of unregulated behaviour (which in *Suicide*, it will be recalled, Durkheim attributed to employers and capitalists primarily). In his lectures on socialism Durkheim considers the issue in relation to workers:

> For it is in vain that one will create privileges for workers which neutralize in part those by employers; in vain will the working day be decreased or even wages legally increased. We will not succeed in pacifying roused appetites because they will acquire new force in the measure they are appeased. There are no limits possible to their requirements. To undertake to appease them by satisfying them is to hope to fill the vessel of the Danaides.[1]

[1] *Socialism*, Collier, 1962, p. 93.

(2) A policy of laissez-faire in which behaviour in the economic sphere is left to market forces is no solution. The interests of individuals when pursued with no restraints other than the free market does not lead to social harmony. This would do nothing to assuage the problem of contractual inequality which needed to be dealt with if one were to talk about fair exchanges in the market.

> Common morality very severely condemns every kind of leonine contract wherein one of the parties is exploited by the other because he is too weak to receive the just reward for his services.[1]

Contract law tends to impinge on the market system such that individuals may receive some protection from the consequences of inequalities of power inherent in a system which reacts in one way to abundance and in another to scarcity.

(3) But could not one argue that in an industrial society industrial interests were coterminous with social interests and planning for social regulation proceed accordingly? This underlies the notion of an administrative society proposed by Saint-Simon. Governmental activity would thus appear to be reduced to a minimum. Such a withering away of the state might suggest anarchy but in fact the original governmental organs would be replaced by the sovereign organization of a supreme council of industry. It is the extension of the administrative method of running banks, insurance companies and business enterprises to society as a whole. In this sense industrial society is to be run industrially. Certainly Saint-Simon saw such a society as voluntaristic and not in need of force to ensure compliance. It is obviously a long way from a laissez-faire system and is intended to provide a moral basis for social life, and, in particular, would regulate property so that resources would be utilized in the most productive manner. Durkheim's objection, however, is that this is a false solution to the problem of the moral integration of the individual in society because moral ends need to be distinguished from economic ends. Does one establish social harmony by maximizing production in order to satisfy the needs of all men as quickly as possible? Durkheim's reply separates him decisively from Saint-Simon:

> Picture the most productive economic organization possible and a distribution of wealth which assures abundance to even

[1] *Division of Labour*, op. cit., p. 386.

the humblest—perhaps such a transformation at the very moment it was constituted, would produce an instant of gratification. But this gratification could only be temporary. For desires, though valued for an instant, will quickly acquire new exigencies . . . even apart from any feeling of envy, excited desires will tend naturally to keep outrunning their goals, for the very reason that there will be nothing before them which stops them.[1]

The cure for anomie therefore is not to subordinate all social activity to economic interests; not even when this is done in the name of morality. Rather, it is society itself which has, so to speak, to be reclothed in moral character for the economic order itself to be integrated into social life (system integration). Without this the integration of the individual into industrial society could not be realized. However, whereas in former times religion had exercised a monitoring moral function, or princely powers had been able to keep industrial activity in check, in nineteenth-century Europe these forces seemed to Durkheim to be largely played out. A key problem therefore remained:

> . . . under the present conditions of social life what moderating functions are necessary and what forces are capable of executing them? The past . . . indicates the direction in which the solution should be sought. What, in fact, were the temporal and spiritual powers that for so long moderated industrial activity? Collective forces. In addition they had the characteristic that individuals acknowledged their superiority, bowed voluntarily before them, did not deny them the right to command. Normally they were imposed not through material violence but through their moral ascendency. This is what accounted for the efficacy of their actions. So, today, as formerly, there are social forces, moral authorities, which must exercise this regulating influence and without which appetites become deranged and economic order disorganized.[2]

(4) Does this lead Durkheim to advocate state regulation in order to provide the sources of collective moral order? To do so would have been quite contrary to his conception of organic solidarity. And he does in fact make his position plain in the matter:

[1] *Socialism*, p. 242.   [2] *ibid.*, pp. 243–4.

A society composed of an infinite number of unorganized individuals, that a hypertrophied State is forced to oppress and contain, constitutes a veritable sociological monstrosity. For collective activity is always too complex to be able to be expressed through the single and unique organ of the State. Moreover, the State is too remote from individuals; its relations with them too external and intermittent to penetrate deeply into individual conscience and socialize them from within. Where the State is the only environment in which man can live communal lives, they inevitably lose contact, become detached, and thus society disintegrates. A nation can be maintained only if, between the State and the individual, there is intercalated a whole series of secondary groups near enough to the individual to attract them strongly in their sphere of action and drag them, in this way, into the general torrent of social life.'[1]

Social reconstruction then, must seek to fill the void between the State and the individual, not by foolish attempts to put the clock back by the revival of archaic and irrelevant social organizations, but by creating new secondary groupings to meet the new circumstances of industrial society.

We are now at a stage where we may note that within the corpus of Durkheim's writings the following propositions are at least involved in solving the problem of anomie:

(*a*) Collective forces are necessary to bind an individual to society and hence prevent the detachment of the individual from society (and hence the egoism of excessive individualism which leads to a sense of the meaninglessness of social life).

(*b*) The state on its own is too removed from the individual to succeed thus in forging the social bonds which are the prerequisite of the individual's social integration.

(*c*) While the state is insufficient in this respect, if it is unchecked by any other collective force it may become despotic in practice. But since in Durkheim's view the very function of the state is to be the liberator of the individual this would result in its moral character being obliterated in the name of tyranny and repression.

(*d*) Appropriate secondary groupings must therefore be

[1] *Division of Labour*, p. 28.

encouraged to protect against the potential tyranny of the State and to foster the social integration of the individual.

(e) However, these secondary groups can themselves become tyrannical, mould the wills and monopolize the lives of their members.

(f) Consequently these secondary groupings must flourish in a context in which they are all subject to an authority (the State) which provides a rule of law for all and which serves to remind each group that they are an interdependent part of a larger whole. Durkheim argues, for example, that the State served to free the child from family despotism, the citizen from feudal domination and the craftsman from guild tyranny, and as such must be considered as essential to the emancipation (social integration) of the individual. A key passage on this occurs in *Professional Ethics and Civic Morals*:

> The State . . . in holding its constituents' society in check . . . prevents them from exerting the repressive influences over the individual that they would otherwise exert. So there is nothing inherently tyrannical about State intervention in the different fields of collective life; on the contrary, it has the object and the effect of alleviating tyrannies that do exist. It will be argued, might not the State in turn become despotic? Undoubtedly, provided there were nothing to counter that trend. In that case as the sole existing collective force, it produces the effects that any collective force not neutralized by any counter force of the same kind would have on individuals. . . . The inference to be drawn from this comment, however, is simply that if that collective force, the State, is to be the liberator of the individual, it has itself need of some counter-balance; it must be restrained by other collective forces, that is by those secondary groups we shall discuss later on.[1]

These points taken together constitute what is perhaps somewhat misleadingly called the corporate solution to the problem of the relationship of the individual to society and in particular to the problem of anomie in modern industrial society.

(5) Durkheim's solution is at once more subtle and more radical than generally realized, and it remains for us to comment on this

[1] op. cit., pp. 62–3.

in relation to his analysis of economic life. It is here that he lays great emphasis upon the importance of establishing occupational and professional associations in trade and industry. In *Professional Ethics and Civic Morals*, for example, he underlines the crucial part of the craft guilds as a source of moral regulation in mediaeval European cities. But as a system of regulation it was static, a prisoner of tradition, unable to cope with innovation and in particular unable to adjust to the advent of large-scale industry. The growth of a national economy led to attempts to exercise direct state control over industry, but the growing diversity and vastness of economic life made this increasingly difficult. What was needed was an intermediate form of organization having a different character from the guild but exercising a form of regulation over social and economic life which could operate on a national scale. He sketches it out in the following way:

> Let us imagine—spread over the whole country—the various industries grouped in separate categories based on similarity and natural affinity. An administrative council, a kind of miniature parliament: nominated by election, would preside over each group. We go on to imagine this council/parliament as having the power, on a scale to be fixed, to regulate whatever concerns the business: relations of employers and employed—conditions of labour—wages and salaries—relations of competitors one with another, and so on . . . and there we have the guild restored, but in an entirely novel form. The establishment of this central organ appointed for the management of the group in general, would in no way exclude the forming of subsidiary and regional organs under its direction and subordinate to it. The general rules to be laid down by it might be made specific and adapted to apply to various parts of the area by industrial boards. . . . In this way, economic life would be organized, regulated and defined, without losing any of its diversity.[1]

Durkheim pursues the matter to claim that such a form of social organization would be neither a static nor over-uniform entity:

[1] op. cit., p. 37.

Their scope and complexity would protect against inertness. They would comprise elements that were too many and too diverse for a fixed uniformity to be feared. The equilibrium of such organization can only be relatively stable and would therefore be in complete harmony with the moral equilibrium of a society with the same character and in no wise rigid. Too many different minds would be at work for new arrangements not to be constantly preparing or, as it were, in a latent state.[1]

In contrasting the mediaeval guild and the proposed occupational associations, Durkheim is essentially building on the sociological distinction between community and association as forms of social order and recognizing that change is an omnipresent feature of societies based upon associations. Consequently, the static moral equilibrium of the guild system should be replaced by the dynamic moral equilibrium of the new corporate structures.

European industrial societies are then portrayed by Durkheim as being in a state of moral crisis. What is required is an institutional order which would provide a new basis of social cohesion, and hence the framework in which individuals would be bound by ties of interests, ideas and feelings. What Durkheim seems to be assuming, as Gouldner has pointed out,[2] is that social interaction provides the basis for moral beliefs to develop spontaneously. In this new system of regulation he envisages employers and employees represented on the governing body (and possibly elected separately given the conflicts of interests between them on many questions), and concerned with such matters as wages, details of the labour contract, working conditions, and the regulation of industrial disputes. It is the creation of such a framework of rules which Durkheim sees as the priority. However, such a reconstruction would also have implications for the system of property ownership. As the owners of the means of production died out Durkheim envisages these new forms of association taking over. The institutions of inherited wealth would be eroded: the professional groups would become, in the economic sphere, the heirs of the family, so to speak. Only through such a process can one move towards an approximation of the just contract. In this sense certainly Durkheim is a radical:

[1] op. cit., p. 38.   [2] See his editorial introduction to *Socialism*, op. cit., p. 26.

As things are, the primary distribution of property is according to birth (institution of inheritance). The next stage is, that property originally distributed in this way is exchanged by contracts. But it is by contracts which, inevitably, are in part unjust as a result of an inherent state of inequality in the contracting parties, because of the institution of inheritance. This fundamental injustice in the right of property can only be eliminated as and when the sole economic inequalities dividing men are those resulting from the inequality of their services. That is why the development of the contractual right entails a whole recasting of the morals of property.'[1]

(6) We have seen now why Durkheim concentrates so much of his attention on economic life: it was in that sphere that acute anomie was most likely to be encountered. One retains the impression that system integration—the orderly regulation of production and consumption in the context of the rest of social life—takes precedence over social integration—the promotion of equality in the external conditions of the struggle. The former, so to speak, begets the latter. The absence of regulation always spells moral anarchy no matter who formally owns the means of production: 'the state of anarchy comes about not from the machinery of labour being in these hands and not in those, but because the activity deriving from it is not regulated'.[2] Durkheim regards his proposals as urgent. System integration will have to be worked for. It must be clearly recognized that economic functions are only a means to an end; they are 'one of the organs of social life . . . and social life is above all a harmonious community of endeavours. Society has no justification if it does not bring a little peace to men—peace in their hearts and peace in their mutual intercourse. If, then, industry can be productive only by disturbing their peace and unleashing warfare, it is not worth the cost.'[3]

But the proposals, though urgent, are reformist not revolutionary. Rather than advocate the destruction of the existing order, he wants to see a series of successive reorganizations. It is not, as he explains in his lectures on socialism, a matter of putting 'a completely new society in the place of the existing one, but of adapting the latter to the new social conditions'.[4]

[1] *Professional Ethics*, p. 215.   [2] *ibid.*, p. 31   [3] *ibid.*, p. 60.
[4] *Socialism*, p. 246.

It remains for us to observe that while Durkheim offers a diagnosis and cure for the disease of acute anomie in industrial societies, mild anomie is treated as endemic. The very notion of 'progress' whether interpreted in system terms of economic growth, or in individual terms of increased opportunity and wider aspirations, implies a sense of dissatisfaction with the status quo. The point is elegantly made in *Suicide* and with this we conclude our exposition of Durkheim's approach:

> . . . Among peoples where progress is and should be rapid, rules restraining individuals must be sufficiently pliable and malleable. If they preserved all the rigidity they possess in primitive societies, evolution thus impeded could not take place promptly enough. But then inevitably, under weaker restraint, desires and ambitions overflow impetuously at certain points. As soon as men are inoculated with the precept that their duty is to progress, it is harder to make them accept resignation; so the number of the malcontent and disquieted is bound to increase. The entire morality of progress and perfection is thus inseparable from a certain amount of anomie.[1]

The social consequences of rapid industrial change posed for Durkheim both a threat and possibility. The threat was a total breakdown of industrial civilization—a retreat to the Hobbesian war of all against all. The possibility was that men might by taking thought reconstruct the social order. For Durkheim this demanded not the imposition of a Leviathan but the encouragement of all the tendencies which promoted organic solidarity. Written into this perspective was a moral concern such that questions of social justice and individual freedom were paramount. To remind men of these questions and to link the solution to the issue of social equality, was Durkheim's own contribution to the social solidarity he desired for industrial societies. Only by developing and encouraging a critical awareness of the threats endemic to industrial society could the built-in anomic tendencies be kept within reasonable, or to use Durkheim's term, 'normal' limits.

[1] *Suicide*, p. 364.

(2) Hobhouse and Liberal Socialism: a Durkheimian Parallel

The English tradition of liberal socialism provides some illuminating parallels to Durkheim's analysis of industrial society. We will look in particular at the work of L. T. Hobhouse. We find there a stress (in part consciously derived from Comte) on the organic nature of society. 'No one element of the social life stands separate from the rest, any more than any one element of the animal body stands separate from the rest.'[1] But, like Durkheim, he rejects the assumptions of classical liberalism that there is a natural harmony between individual and society such that the unchecked pursuit of enlightened self-interest produces behaviour that promotes the public good. This classical view of liberalism was itself a response to the authoritarian claims of Church and State and against which the inalienable claims of men to their natural rights were proclaimed (for example, liberty, property, security and resistance to oppression as in the 1789 Declaration of the Assembly). In itself, this position does not logically imply that there should be no public control of individual behaviour. The possibility that individuals might come into conflict with one another as they pursued their own ends was indeed the basis of social contract theories about the origin and nature of society: a measure of mutual restraint was the basis of the effective liberty of the individual. However, Hobhouse observes:

> ... In the course of the eighteenth century, and particularly in the economic sphere, there arose a view that the conflict of wills is based on misunderstanding and ignorance, and that its mischiefs are accentuated by governmental repression. At bottom there is a natural harmony of interests. Maintain external order, suppress violence, assure men in the possession of their property, and enforce the fulfilment of contracts, and the rest will go of itself. Each man will be guided by self-interest, but interests will lead him along the lines of greatest productivity. If all artificial barriers are removed he will find

[1] *Liberalism*, Oxford University Press, 1964, p. 41.

the occupation which best suits his capacities, and this will be the occupation in which he will be most productive, and therefore, socially, most valuable. He will have to sell his goods to a willing purchaser, therefore he must devote himself to the production of things which others need, things, therefore, of social value. He will, by preference, make that for which he can obtain the highest price, and this will be that for which, at the particular time and place and in relation to his particular capacities, there is the greatest need. He will, again, find the employer who will pay him best, and that will be the employer to whom he can do the best service. . . . This harmony might require a certain amount of education and enlightenment to make it effective. . . . Government must hold the ring and leave it for individuals to play out the game.[1]

What is here portrayed is a functionalist view of the social and economic order, of a society regulated by the celebrated invisible hand. But did not this order rest upon considerable inequalities of property ownership? The sanguine answer was yes, but if that is the way a system is designed to work, there is no more to be said. We do find, it is true, a sense of a moral dilemma in Malthus' *First Essay on Population*, where he concedes that social institutions which promote inequalities of wealth are bad. But he is not satisfied that governmental action to redress such inequalities is socially desirable, and observes rather: 'Perhaps the generous system of perfect liberty, adopted by Dr Adam Smith and the French Economists, would be ill exchanged for any system of restraint.'

For Hobhouse, however, the opposition between liberty and restraint was a false antithesis. The absence of restraint can permit the infliction of injury by the strong on the weak, and the operation of forced bargains. Such bargains necessarily lack the character of true consent. What is needed is the promotion of conditions in which parties entering into bargaining situations do so on a basis of parity. For this reason, for example, Hobhouse firmly supported the extension of protective legislation for the manual labourer in the field of industrial relations.[2] And, like Durkheim, and, indeed, his distinguished English liberal pre-

[1] *Liberalism*, Oxford University Press, 1964, p. 34.
[2] See, for example, *Liberalism*, p. 51.

decessor J. S. Mill, he questions the institution of property inheritance on identical grounds:

> Is there not something radically wrong with an economic system under which through the laws of inheritance and bequest vast inequalities are perpetuated? . . . Wealth, I would contend, has a social as well as a personal basis. Some forms of wealth, such as ground rents in and about cities, are substantially the creation of society, and it is only through the misfeasance of government in times past that such wealth has been allowed to fall into private hands. Other great sources of wealth are found in financial and speculative operations, often of distinctly anti-social tendency and possible only through the defective organization of our economy. . . . Through the principle of inheritance, property so accumulated is handed on: and the result is that while there is a small class born to the inheritance of a share in the material benefits of civilization, there is a far larger class which can say 'naked we enter, naked we leave'. The system as a whole . . . requires revision . . . the ground problem in economics is not to destroy property, but to restore the social conception of property to its right place under conditions suitable to modern needs. . . . This is to be done by distinguishing the social from the individual factors in wealth, by bringing the elements of social wealth into the public coffers, and by holding it at the disposal of society to administer to the prime needs of its members.[1]

We have already seen that for Durkheim the economic order was treated as a sphere which must be subordinated to the wider social order: this is implied in his treatment of morality. In Hobhouse a similar position is maintained. Certainly justice in the economic sphere is not to be treated as deriving from power relations in the economic order. Restraint is necessary. To the contention that wages, to take a crucial example, should be determined by the higgling of the market, Hobhouse replies that this:

> from the social point of view is a council of despair. It holds the door permanently open to a quarrel whenever a change of

---

[1] *Liberalism*, pp. 97–8, very similar views are also very powerfully expressed in R. H. Tawney, *Equality*, George Allen & Unwin, 1964.

conditions occurs, or whenever a combination of employers or employed sees a favourable opportunity for a move, and it leaves each quarrel to be determined by the strength of the parties at the moment, without reference to the permanent needs of industry.[1]

This lack of adequate regulation leads Hobhouse to comment that this is a basis for industrial warfare and as such is socially and commercially destructive. The collective moral solution of Hobhouse has a clear Durkheimian ring about it:

> The opportunities of work and the remuneration for work are determined by a complex mass of social forces which no individual, certainly no individual workman, can shape. They can be controlled, if at all, by the organized action of the community, and therefore, by a just apportionment of responsibility, it is for the community to deal with them.[2]

The spirit of Hobhouse's approach is again akin to Durkheim's in its reformist character. The construction of Utopias, he argues, is not a sound method of social science. Liberal socialists must recognize that what they have to do is to forsake the Utopian visions which have an illusory simplicity about them and present instead 'not a system to be formulated as a whole for our present arrangements but a principle to guide statesmanship in the practical work of reforming what is amiss and developing what is good in the actual fabric of industry'.[3] Hence, to revert to the question of wage settlements as a case in point, Hobhouse argues that modes of peaceful adjustment are in principle possible, and, in the English context, he commended the role of Trade Boards, Whitley Councils and the like, which could provide a basis for making impartial decisions. They could, through a trial and error process, arrive at solutions which would prove generally acceptable to employers and employees. A body of case law would in time emerge:

> Naturally controversies arise, but in a representative body of employers and workers the circumstances which have led to

[1] 'Industry and the State' in *Sociology and Philosophy*, Bell, 1966, p. 219.
[2] *Liberalism*, op. cit., pp. 86–7. [3] *ibid*., p. 89.

the existing grading are pretty well known, and it is generally held that graded rates should be of uniform application, while if there are reasons for any local exception or if there are arguments for a change, they are well within the powers of the technical men to appreciate.[1]

The individual industry boards—with employers, employees, and independent members—Hobhouse saw as being linked into a central board, similarly composed to take into account interindustry wage movements and to advise and recommend on wage settlements accordingly. Now although Hobhouse does not envisage a political extension of these boards along the lines Durkheim appeared to have in mind ultimately for occupational associations, their functions clearly overlap in so far as they are regarded as important regulatory agencies, having a collective force which is derived from a social morality of 'fairness' and applied to the industrial sphere. It is, one may say, the moral alternative to industrial warfare as a basis for wage settlements. Elsewhere, and usefully, Hobhouse distinguishes between attempts at constructing impartial control systems regulating matters such as wages, hours and working conditions, and the executive direction of industry. Here he casts doubt upon private enterprise management being innately superior to other forms of management, and suggests other modes—such as joint boards of consumers and producers, co-operative associations, municipal enterprise: 'according to the nature of the industry, and the relative efficiency for varying purposes of which various forms of organizations prove themselves capable'.[2]

Finally, in respect of Hobhouse's contributions, we may remark that for him, as for Durkheim, the organic character of modern industrial society is treated as an ideal to be more nearly obtained than hitherto. There can in practice be uneven developments in different spheres of social life. But 'by keeping to the conception of harmony as our clue we constantly define the rights of the individual in terms of the common good, and think of the common good in terms of the welfare of all the individuals that constitute a society'.[3] Hence the social integration of the individual into society

[1] 'Industry and the State', op. cit., pp. 219–20.
[2] *The Elements of Social Justice*, Allen & Unwin, 1921, p. 184.
[3] *Liberalism*, op. cit., pp. 108–9. But see below pp. 199–201 for a more sceptical view of the concept of harmony by a modern liberal R. Dahrendorf.

was partly and to an important degree to be related to the appropriate social regulation of economic life. But this, in the end, could not be separated from the rest of social life. Thus if it be true that 'people are not fully free in their political capacity when they are subject industrially to conditions which take the life and heart out of them',[1] to reform those same conditions was not in itself enough. As with Durkheim, other forms of secondary organizations between man and State must also be in good working order.

> The development of social interest—and that is democracy—depends not only on adult suffrage and the supremacy of the elected legislature, but on all the intermediate organizations which link the individual to the whole. This is one among the reasons why devolution and the revival of local government, at present crushed in this country by a centralized bureaucracy, are of the essence of democratic progress.[2]

Much of Hobhouse's work was indeed bound up with the concept of citizenship and with the movement towards the equality of rights of all citizens, despite differences of function, in liberal democratic societies. This, as we have already seen, had political and economic dimensions. Hobhouse recognized that these developments could be uneven in actual societies. Other writers have put the matter differently to suggest that development in one sphere may even be at the expense of development in the other. Alfred Marshall wrote of the concept of citizenship with its implications of equality before the law and educational opportunities for all, co-existing with economic inequalities.[3] This provided the point of departure for T. H. Marshall's study *Citizenship and Social Class* delivered as a series of lectures in 1949:

> Is it still true that basic equality, when enriched in substance and embodied in the formal rights of citizenship, is consistent with the inequalities of social class? I shall suggest that our society (Britain) today assumes that the two are still compatible,

---

[1] *Liberalism*, op. cit., p. 126.   [2] op. cit., pp. 118–19.
[3] A. Marshall, *The Future of the Working Classes*, Thomas Tofts, 1873.
[4] Reprinted in T. H. Marshall, *Sociology at the Crossroads*, Heinemann, 1963.

so much so that citizenship has itself become, in certain respects, the architect of legitimate social inequality.[1]

Marshall pointed out that citizenship could be regarded as a status marking full membership of a community: with an implication of equal rights and duties. He noted that the growth and extension of citizenship could be traced in England from the late seventeenth century. The paradox was that this coincided with the growth of capitalism which was manifestly a system of inequality. This was partly explained by the fact that citizenship was treated in terms of individual civil rights: these were necessary for a competitive market economy to work at all with their emphasis on individual freedoms. They did nothing in themselves to remove inequalities of power:

> For modern contract is essentially an agreement between men who are free and equal in status, though not necessarily in power. Status was not eliminated from the social system. Differential status associated with class, function and family, was replaced by the simple uniform status of citizenship, which provided the foundation of equality on which the structure of inequality could be built.[2]

Citizenship, however, for Marshall also has a political and a social element: one referring to the right to participate in the exercise of political power and the other to rights of economic and social security and 'to live the life of a civilized being according to the standards prevailing in the society'.[3]

Part of Marshall's thesis was that the trade union movement began to exercise civil rights collectively on behalf of their members and that, moreover, this was used as a basis for claiming certain social rights as their social and economic status was raised. Indeed, before the working class became fully aware of its growing political power through the franchise, trade unionism in Marshall's view 'created a secondary system of industrial citizenship parallel with and supplementary to the system of political citizenship'.[4] Broadly then there is, among the working class, a movement from individual to collective action in civil rights portrayed which extends to political and social developments of

[1] op. cit., p. 73.   [2] op. cit., p. 91.   [3] op. cit., p. 74.   [4] op. cit., p. 98.

citizenship. More recently Reinhard Bendix has applied Marshall's thesis to Western European societies undergoing the process of industrialization:

> In Europe the rising awareness of the working class expresses above all an experience of *political alienation* that is a sense of not having a recognized position in the civic community or of not having a civic community in which to participate. . . . Rather than engage in a millenarian quest for a new social order, the recently politicized masses protest against their second class citizenship, demanding the right of participation on terms of equality in the political community of the nation-state. If this is a correct assessment of the impulses and half-articulated longings characteristic of much popular agitation among lower classes in Western Europe, then we have a clue to the decline of socialism. For the civic position of these classes is no longer a pre-eminent issue in societies in which the equality of citizenship has been institutionalized successfully.[1]

There are two general points that we may here note. First incorporation of the working class into the nation is somewhat ambiguous. This is because, in the English case, the trade unions as the fourth estate attacked the social system from the outside, so to speak, in asserting their social rights, but gradually the leadership has become involved in a co-operative relationship with successive governments. As Marshall points out, however, so far as the rank and file trade unionist is concerned there may be a conflict between the welfare of the wider community and economic interest and, in practice, the status rights of industrial citizenship may be asserted over those of political citizenship. Some unofficial strikes in breach of agreements and against the union leadership could, in Marshall's view, be explained in this way. Secondly, Marshall observed that while British society could be characterized as a stable democracy this did not diminish the fact that marked economic inequalities co-existed with certain movements towards greater social equality. And he concluded significantly: 'It may be that some of the conflicts within our social system are becoming too sharp for the compromise to

[1] R. Bendix, *Nation-building and Citizenship*, Anchor Books, 1969, pp. 88–9.

achieve its purpose much longer.'[1] Our further discussion of industrial relations, later in this chapter, will bear upon this issue from the perspective of the anomie theme. We will, in addition, take up the question of social integration again in the concluding part of this book.

(3) *The Breakdown of Community as a Form of Anomie*

A more than usually clear example of community breakdown as a product of specific change in the industrial sphere, is located in Cottrell's American study *Death by Dieselisation: a Case Study in the Reaction to Technological Change.*[2] Caliente had been built as a service point for steam engines. The advent of the diesel engine meant that the community was no longer necessary for the railroad company. This meant in practice that the community simply shrivelled up. As Cottrell points out, there is more to such an event than the balancing of the railway company's profit and loss account. The human consequences affected employees who lost their seniority rights vested in the local railway shops, home owners whose property dropped sharply in value, merchants whose businesses and property were devalued, and the bondholders who could scarcely foreclose on a dying town:

> In a word, those pay who are, by traditional American standards, *most moral*. Those who have raised children see friendships broken and neighbourhoods disintegrated. Those who built their personalities into the structure of the community watch their work destroyed. . . . In short, *good citizens* who assumed family and community responsibility are the greatest losers. . . . The people of Caliente are asked to accept as 'normal' this strange inversion of their expectations.[3]

[1] op. cit., p. 127.
[2] In *ASR* XVI (1951) and reprinted in N. J. Smelser (ed.), *Readings on Economic Sociology*, Prentice-Hall, 1965.
[3] op. cit., p. 238.

Although various forms of pressure group activity were undertaken by the community, their attempts to influence either the railroad company or the government were unsuccessful, and what Cottrell is emphasizing is the fact that communities are necessarily dependent upon outside groups or interests which may subordinate the interests of the local community to wider economic interests. Virtue is penalized rather than rewarded: the citizens discover that market forces determine their destiny. Here, then, one encounters a pervasive sense of injustice and demoralization—the marks of anomie—as the community succumbs to technological change which it neither invented nor could control. Cottrell points out that the case was by no means unique in American railroad history and it finds its parallels in all industrial societies, notably in communities that are based on single industries. The mining industry in the UK provides a notable contemporary case in point to which the depopulated villages of South Wales and Durham bear witness.

A very well known case of community breakdown as a result of changes in the economic sphere is documented in Warner and Low's study *The Social System of a Modern Factory*.[1] It documents and seeks to account for a strike which hit the shoe industry of Yankee City during the depression in 1933. When one describes the story as a 'breakdown of community' one is referring to a number of interrelated changes. These are schematically portrayed by Warner and Low in a series of phases, and relate to changes in technology, the division of labour, the form of ownership and control, producer-consumer relations, worker relations, and the structure of economic relations. (See chart, p. 102.)

One could observe:
1. A change from single hand tools eventually to machine-based mass production methods.
2. A movement away from individual and sometimes highly skilled work, to unskilled, routinized labour.
3. The replacement of local control by outside ownership and control.
4. A change in the market from one based upon local or regional outlets, to one in which local factories simply become one source of supply for a chain of shoe stores.

[1] W. Lloyd Warner and J. O. Low, *The Social System of a Modern Factory. The Strike: A Social Analysis*, Yale University Press, 1957.

The History of the Differentiation of the Yankee City Shoe Industry

	Technology	Form of Division of Labour	Form of Ownership and Control	Producer-Consumer Relations	Worker Relations	Structure of Economic Relations
IV The Present (1920-1945)	Machine Tools mass production, assembly line methods	Nearly all jobs low skilled; a very large number of routinized jobs	Outside ownership and control of the factory (tools leased)	Very few retail outlets; factory merely one source of supply for a chain of shoe stores	Rise of industrial unions, state supervised . . . no (or weak) unions	Center of dominance New York. Very complex financial producer and retail structure. Local factory not important in it
III Late Intermediate Period (approximately) to World War I	Machine Tools machines predominate; beginning of mass production through use of the machine (McKay)	A central factory with machines; still high degree of skill in many jobs	First small, and later, large *local* men of wealth own or lease the tools, and machines	National market and local capitalist; many outlets	Craft and apprenticeship (St Crispin's Union)	Center of dominance local factory; complex hierarchy in local factory system
II Early Intermediate Period (approximately) to the Civil War	Machine Tools few machines first application (Elias Howe, etc.)	One man assigns highly skilled jobs to few men; highly skilled craftsmen ("letting-out" system)	Small, locally controlled manufacturers; tools still owned by workers, materials controlled by "owner"	Owner and salesmen to the consumer re-regional market	Informed, apprenticeship and craft relations	Simple economic no longer kinship; worker subordinate to manager
	Hand Tools increasing specialization and accumulation of hand tools	Specialization among several families; a few highly skilled jobs	*Local Control* not all shoemakers need own all tools; beginning of specialization	Local buyer from several producer families sells products (no central factory)	Kinship and neighbours among workers	Semi-economic but also kinship and neighbourliness
I The Beginning (early 1600's)	Hand Tools few, basic, and simple	All productive skills in the family, including making of shoes; a few cobblers for the local market	*Local Control* skills, tools and materials owned and controlled by each family; or by the local cobbler	The family produces and consumes shoes and most other products	Largely kinship and family relations among workers	Very simply non-economic; the immediate family

Source: *The Social System of the Modern Factory*, op. cit.

5. The replacement of kinship and craft based worker relations by industrial unions.
6. The transformation of the former structure of economic relations based on localized elites and status distinction by the fact that the local factories become absorbed into a vast commercial complex dominated by New York City.

What is being described is the breakdown of community as a total society. Whatever limitations the local system had placed on the workers (and one always runs the risk of putting a fair dose of nostalgia into the community concept), there were many accompanying disadvantages with the new situation so far as the workers were concerned:

> The workers of Yankee City were able to strike, maintain their solidarity ... and in a sense flee to the protection of the unions because the disappearance of craftsmanship and the decreasing opportunities for social mobility had made them more alike with common problems and common hostilities against management. The craft differences had been wiped out, and occupation mobility in the craft hierarchy and secondarily social mobility in the community had been stopped. The workers felt even more alike and were increasingly motivated to act together because their new occupational status had contributed to their downward orientation in the community.[1]

Warner and Low are explicit in acknowledging their intellectual debt to Durkheim's *Division of Labour*.[2] What is particularly noteworthy however is that their confidence in advanced industrial societies achieving organic solidarity is very small indeed. On the contrary, they see what Durkheim had defined as inequalities in the external condition of the struggle as widening. It is this conviction which gives a strong and perhaps surprising Marxist colour to their general analysis:

> Despite the great changes in our economic and social order the American class system continues.... There are many who believe the increasing power exercised by the political, economic and associational organizations of the workers will result in an

[1] op. cit., p. 171. [2] See *Social System*, p. 54.

equalitarian society. The present writers do not share their opinion. We believe that, with the increasing complexity of our society and the greater development of our relations with other great world societies, the hierarchical forms operating in our society will increase their power.[1]

This point is made notwithstanding the acknowledgement that material standards will improve for the lower classes. It is the phenomenon of bureaucratization linked to monopoly capitalism that are of central concern and of which the community breakdown at Yankee City is seen as symptomatic. The basic instability in modern industrial societies is summed up in the observation that 'no social order extending itself vertically and horizontally, while systematising and rationalizing the actions and relations of its members, can ever achieve any kind of permanent equilibrium.'[2] Warner and Low, however, share with Durkheim the conviction that the insatiable demands of the economic order need to be held in check by the adequate development of other powerful institutional orders. This is the significance of the following comment:

> The extension both horizontally and vertically of the international economic institutions over men and a greater variety of their activities will only be seen when the economic institutions are complemented by church, associational and political hierarchies. The development of these other institutions will permit them to act internationally as counterforces to achieve a balance of power between them and the economic order. . . .[3]

The great emphasis here, therefore, is on system integration through the existence of counteracting power in different institutional orders. The question of social integration is not ignored either, and Warner and Low stress the importance of equal educational opportunity to increase the individual's opportunity for social mobility. This, they argue, would decrease the chance for revolutionary outbreaks expressing frustrated aspirations. They took the view then, however, that the class system was becoming less open, and that mobility was increasingly difficult for those at the bottom of the social heap. We may recall that the Yankee City studies related to data collected in the 1930s–40s.

[1] *Social System*, p. 188.   [2] *ibid.*, p. 193.   [3] *ibid.*, p. 196.

But the issue of inequalities of opportunity remains. So, for example, Blau and Duncan's comprehensive study, *The American Occupational Structure*[1] concludes that men who begin their working lives in the working class end up in occupational positions that differ little from their fathers and that the influence of social origins on career beginnings has not changed at all in the last forty years.[2] The point is made that 'superior family origins increase a son's chances of obtaining superior occupational status in the United States in large part because they help him to obtain a better education'.[3]

The 'breakdown of community' theme draws attention to the normlessness generated by certain forms of industrial change and development. It is not, however, only a question of small-scale communities disintegrating but of their being replaced by or transformed into large cities and the anomic tendencies that are thereby entailed.

A good deal of sociological literature focuses upon the anomic character of city life. To Durkheim, for example, the city represented a social context within which collective forces over the individual must necessarily be weakened. They are areas characterized by considerable migratory movements. The authority of the older generation upon the younger is weakened by geographical mobility: the city is essentially an arena for innovation with all the benefits and costs which that may entail:

> ... nowhere have the traditions less sway over minds. Indeed, great cities are the uncontested homes of progress; It is in them that ideas, fashions, customs, new needs are elaborated and then spread all over the rest of the country. When society changes, it is generally after them, and in imitation. Temperaments are so mobile that everything that comes from the past is somewhat suspect. On the contrary, innovations, whatever they may be, enjoy a prestige there almost equal to the one the customs of ancestors formerly enjoyed. Minds are naturally there oriented to the future. Consequently, life is there transformed with extraordinary rapidity: beliefs, tastes, fashions, are in perpetual evolution. No ground is more favourable to evolutions of all sorts. That is because the collective life cannot

---

[1] John Wiley & Sons, Inc., 1967.   [2] See p. 424.   [3] *ibid.*, p. 430.

have continuity there, where different layers of social units, summoned to replace one another, are discontinuous.[1]

What one should recognize about a passage like this is that it is certainly not a polemic against the city, nor does it have any touch of agrarian mysticism about it in the manner, say, of Spengler or even Tönnies. Rather it is a situation fraught with possibilities as well as dangers—a development away from mechanical solidarity towards organic solidarity in which individual personalities may develop. The important thing, rather, is that a new form of social solidarity exists to hold in check, so far as possible, the tendencies to anomie which are inherent in the situation and which threaten to crush the new-found opportunities for individuality and autonomy. The double-edged character of city life is similarly noted by Merton:

> ... much the same conditions of urban life—a degree of emancipation from localistic control, harsh competitiveness, a measure of impersonality—underlie both aberrant behaviour and human accomplishment. The free ranging intellect exercising its muscles to the full, thrives in an atmosphere of relative autonomy in the midst of plenty of people, just as in another direction, roving muggers and thugs thrive in an atmosphere that provides room for individual movement, not easily subjected to social control.[2]

Not surprisingly, in Western Europe and the United States the rapid growth and extension of city life in the nineteenth and twentieth centuries has led social scientists to chart these developments. The growth of sociology in the United States can, indeed, be seen in large measure as an attempt to grapple with the many social problems created by urbanization.

We would recall in particular Elton Mayo and his collaborators whose writings explicitly reflect an interest in solving the problem of anomie in an industrial society and who, in terms of his own immediate experience, was clearly much influenced by the 'melting-pot' Chicago of the 1920s and '30s. One overriding conceptual distinction in his writings relates to his classification of societies

---

[1] *Division of Labour*, op. cit., p. 296.
[2] *Anomie, Anomia and Social Interaction*, op. cit., p. 224.

into *established* and *adaptive*. With engaging simplicity he denotes all pre-industrial societies as *established*—having set customs, rituals and moral rules. Each member of the established society knows his place, and in that knowledge obtains a sense of personal security and emotional well-being. An established society was above all an ordered society in which each member collaborated spontaneously to ensure its maintenance. Such societies were obliterated by the industrial revolution. The *adaptive* society was essentially characterized by continual technological change. The technical skills, which the growth of science and industry represented, were not in Mayo's view matched by a growth in social skills; hence rapid social change, to date, has spelt disorder and social disintegration. This was notably reflected, he maintained, in the break-up of primary group life. What then could be done? First, we should recall that man is a social animal and finds important sources of personal and emotional satisfaction in the membership of small groups. In an industrial society this obviously included the work group. With this in mind Mayo wrote forcibly against the 'rabble hypothesis': the view that society is composed of individuals each logically acting out of self-preservation or self-interest. The essence of society is co-operation between men, and where there is no co-operation there is individual isolation, rootlessness and disorder. In industry, for example, this could be reflected in high absenteeism and sickness rates, high labour turnover and, in general, a lowering of efficiency arising from a weak commitment to work. What one should do, therefore, is to recognize the reality of primary group life in the work place and do everything possible to sustain and develop it.

But to what end? On this Mayo appeared to have no doubts. A persistent problem of management was to organize sustained co-operation of its employment force in the face of continual change. By recognizing the work group as a social context in which employees can fulfill their desire for co-operative activity one fostered team work (for example, by showing approval of work done, by showing individuals how their work fitted into the whole, by allowing individuals some freedom of choice over whom they worked with) in the service of management objectives. It is the manager, then, assisted by the consultant social scientist, who must exhibit the social skills necessary to promote such social integration.

We may recall that in the Hawthorne investigations, following the Relay Assembly Test Room studies, Roethlisberger and Dickson observed:

> What impressed management most . . . were the stores of latent energy and productive co-operation which clearly could be obtained from its working force under the right conditions. And among the factors making for these conditions the attitudes of the employees stood out as being of predominant importance.[1]

We are dealing then, in principle, with a managerially-induced form of co-operation. However, even though this was a case of a work group conforming quite closely to management's expectations, it appeared not to have been established without the use of sanctions. In the seventh period of the Relay Assembly Test Room experiment, two of the five assemblers were held to constitute a 'personnel problem'. 'It had been apparent for quite a while that these operators were not displaying that "wholehearted co-operation" desired by the investigators.'[2] This was manifested in what the supervisors felt to be excessive talking. The two girls displayed hostile attitudes towards authority and reduced their output, and this led directly to them being replaced by two others who proved to be more co-operative. Roethlisberger and Dickson appear to be slightly apologetic about this in reviewing the incident, arguing that the investigators wrongly tried to hold 'the spirit of co-operation' as a constant in the experiment. Rather, they should have looked at the causes of the lack of co-operation, the implication being that they should have sought for ways of promoting the appropriate change in the deviant employees' attitudes.

It might be thought that—given this experience, together with the evidence of output restriction in the Mica Splitting group and the Bank Wiring group—the integration of the individual into the work group was at the expense of managements' interests. Mayo, however, tended to treat such differences as a challenge to management to develop their social skills. 'For the larger and more complex the institution the more dependent is it upon the

---

[1] *Management and the Worker*, John Wiley & Sons Inc., 1964, p. 185.
[2] *ibid.*, p. 53.

wholehearted co-operation of every member of the group.'[1]
Better human relations through effective communication was to become the slogan for this approach to social integration.

Mayo, like Durkheim, argued that in an industrial society the state could not of itself serve as an integrating institution, but there, we may suggest, most of the similarity between the two ends. There is, for example, no general discussion of the role of other secondary organizations. And, as many critics have pointed out, it is difficult to avoid the conclusion that social integration is defined in terms of employee manipulation by a managerial élite in the company. Because the overall goal of managerially defined efficiency is taken as a guiding light, conflicts of interest are treated as pathological, whereas in Durkheim it was clearly recognized that conflict could not be eliminated from social life and one should, therefore, focus attention on conflict regulation. Indeed attempts to abolish conflict were treated by Durkheim as repressive. The essence of social reconstruction for Mayo, however, involved curing the disease of conflict:

> The administrator of the future must be able to understand the human and social facts for what they are, unfettered by his own emotion and prejudice. He cannot achieve this ability except by careful training that must include knowledge of relevant technical skills, of the systematic ordering of operations, and of the organization of co-operation.[2]

The most well-known expression of the therapeutic approach to induce collaboration was the counselling programme at Western Electric. Skilled interviewers were provided to help employees talk through their problems and difficulties, so that they might find satisfaction in their work life, and, of course, improve their commitment to work, perhaps aided by a better understanding of company policies.

This approach to social integration through the development of plant harmony has been castigated by its critics as 'managerial sociology'. Daniel Bell, for example, maintains:

---
[1] *The Social Problems of an Industrial Civilization*, Routledge & Kegan Paul, 1949, p. 62.
[2] *ibid.*, p. 43.

The gravest charge that can be levelled against these researchers is that they uncritically accept industry's own conception of workers as a *means* to be manipulated or adjusted to impersonal ends. The belief in man as an end in himself has been ground under by the machine, and the social science of the factory researchers is not a science of man but a cow sociology.[1]

Certainly the union as an alternative source of worker integration is not stressed. As a cure for worker isolation unions were a poor second best. If management was doing its job properly in fostering loyal collaboration through improved communications, developing welfare policies and the like, union membership was unnecessary. And a union which was organized to combat managerial policies was treated as a source of disorder.

The plant society is therefore treated as the primary context of social integration, and a company as an organization acts as a buffer against the anomic tendencies of city life. However, in Roethlisberger and Dickson's account of the Hawthorne investigations, there are traces of another perspective which Mayo himself seems to have ignored. In their discussion of supervisors' job satisfaction they offer the following generalizations:

> The more impoverished the social reality for the supervisor the greater are his feelings of insecurity and the greater are his demands for recognition and security. For some supervisors the company is father, mother, society and state all rolled into one, and strivings for success are a compensation for lack of normal and adequate personal interrelations. Such supervisors as these are attempting to substitute the company for the wider social reality.[2]

And they go on to make the interesting observation that the supervisors who seemed most contented were those with active political and social affiliations and foreign-born supervisors having residential ties with their ethnic groups. All this of course would suggest a different model of social integration which could not ignore life outside the company and, indeed, has pluralist implications which, we would suggest, are more Durkheimian in character

[1] D. Bell, *Adjusting Men to Machines*, Commentary, Vol. 3, 1947, p. 88.
[2] op. cit., p. 372.

than Mayo's own position. But it is not a question which is systematically pursued.

There is no discussion in Mayo of inequalities in the external conditions of the struggle as in Durkheim, and in consequence no attention is paid to the existing system of property relations and the distribution of wealth as subjects requiring attention. In this sense, Mayo is concerned with shoring up the existing system rather than reconstructing it. In this important respect his proclaimed allegiance to the Durkheimian tradition is somewhat misleading. And we may further recognize that there is nothing really approximating to the occupational association which figures so crucially as a regulating group in Durkheim's analysis. Such an association could presumably recognize the usefulness of mobility of labour between firms, both for economic reasons and also to enable a person to reach a higher level of responsibility and attainment. In Mayo great emphasis is laid on reducing labour turnover. Certainly one can well understand how high labour turnover might constitute a managerial problem. At the same time, in Durkheim's terms, to reduce it by managerial techniques might prevent individuals from occupying a place in the social framework compatible with their faculties—notwithstanding an employee development programme within the company. Finally, we may note that there is in Mayo no discussion of system integration in the sense of modes of regulating production and consumption to smooth out the imbalances which promote inflationary or deflationary movements in economic life. Rather, for Mayo, it is a question of managers deploying social skills to ensure the adaptation and survival of the primary group in the face of change. A macro-theory of the causes of change or a discussion of the institutional mechanisms whereby it may be regulated is not propounded. In particular, there is no discussion of the institutional framework of industrial relations and collective bargaining. And it is to this dimension of economic life that we now turn.

(4) Social Order, Anomie and Industrial Relations

One of the most explicit attempts to analyse industrial relations within a Durkheimian perspective is to be found in Fox and Flanders' recent essay 'The Reform of Collective Bargaining: from Donovan to Durkheim'.[1] As the title of the paper implies, their analysis is related to the contemporary British industrial relations scene. We will attempt here to review the substance and, as we see it, the implications of their argument. 'Industrial relations' is portrayed as being essentially concerned with the making and administration of rules regulating employment relations, and 'collective bargaining' is seen as the principal norm-creating institution in industrial relations. It is 'the principal method evolved in industrial societies for the creation of viable and adaptive normative systems to keep manifest conflict in employment relations within socially tolerable bounds. This it has done because the rules it produces, as expressed in collective agreements and in unwritten understandings, are supported by a sufficiently high degree of consensus among those whose interests are most affected by their application.'[2] Nevertheless, for various reasons, collective bargaining may not always prove to be an adequate mode of conflict regulation. Where this is so, a state of disorder or anomie is postulated. The two key questions posed by Fox and Flanders are: first, what causes a breakdown in the system of regulative norms?—and second, what is to be done when such a breakdown is diagnosed?

The response to the first question rests on a fairly complex analysis of the kinds of interaction which may exist between the normative aspirations of a group and the existing system of normative regulation. Four major sources of disorder are postulated:

(i) when one group against the resistance of another attempts to

---

[1] In *British Journal of Industrial Relations*, VII, 1969, reprinted in A. Flanders, *Management and the Unions: the Theory and Reform of Industrial Relations*, Faber & Faber, 1970. Page references are cited from the *Journal*.
[2] *ibid.*, p. 160.

change procedural norms on which industrial relations are based,
(ii) when a group likewise attempts to change substantive norms. These essentially relate to wages and conditions of employment,
(iii) when one or more groups has normative aspirations in areas where no normative regulations exist. Conflicts of interests and values take place without any frame of reference to which conduct might be related,
(iv) where a progressive fragmentation and breakdown of existing regulative systems occurs.

When disorder of the second and third variants occurs more frequently and extensively in the industrial relations sphere, Fox and Flanders suggest that this creates the conditions in which the fourth variant may appear. We are here presented with an amplification thesis of disorder with the fourth variation being first produced by other structural deficiencies and then spawning its own disintegration. The powerful impose their own norms and create their own solutions, those without power experience the frustrations of unsatisfied aspirations. And it is the fourth source of disorder which is treated as a phenomenon of 'special gravity and intractability':

> The other three demand concessions and compromise on the part of resisting groups, possibly on issues which they have regarded as matters of principle, but order can be re-established by a readjustment or extension of existing normative systems. But when the progressive and arbitrary fragmentation of the system has passed a certain point, nothing short of their wholesale reconstruction can remove the source of disorder. In general it can be argued that the more relatively numerous the normative systems regulating employment relations, the greater the problems of social order, since the task of finding the requisite measure of integration among the various systems becomes increasingly difficult to solve.[1]

The 'profound and serious consequences' which this can have for the whole social order are then indicated:

[1] *British Journal of Industrial Relations*, VII, 1969, p. 162.

The proliferation of norm creating groups and the resulting multiplicity of normative systems may produce a degree of disorder which is felt to impede and imperil vital functions of social life and government. In industrial relations the economic consequences are not confined to strikes and other dislocations of productive process. The loss of integration and predictability is also expressed in such things as chaotic pay differentials and uncontrolled movements of earnings and labour costs. And the political consequences are decidedly no less important. Growing disorder may threaten the government's ability to govern and starts to generate strong popular demands for authoritarian State intervention to restore order.[1]

Since this is a situation which Fox and Flanders broadly believe characterizes the British industrial relations scene at the present time much of their analysis is centred on this. However, the attempt is also made to look at the situation in historical perspective, in an attempt to account for the process of normative disintegration. What is there suggested, in fact, is that two forms of fragmentation may affect the system of normative regulation: deflationary and inflationary. The first appeared in periods of high unemployment and the second in periods of low unemployment. The historical thesis propounded is that just prior to the First World War Britain was moving into a position of inflationary fragmentation which in institutional terms was never solved. The First World War intervened. The inter-war years exemplified the fact that the unsolved problems of constructing a mutually acceptable system of industrial relations led to disorder—either in overt management-union conflict or, as the depression deepened, in imposed settlements by employers on workers which they were powerless to resist, but which left a legacy of great bitterness. The Second World War stood as something of a truce period although clearly there was a recovery of trade union influence, but the post-war period reaped the harvest of the unsolved problems arising from inflationary fragmentation. The aspirations of workers in this period are seen as revived, extended and imposed by virtue of the reality of shop floor power on individual employers. The picture drawn is of inflation creating further inflation as a

[1] *British Journal of Industrial Relations*, VII, 1969, pp. 162–3.

result of a process of unrestrained competition. It is reflected in a number of mutually interacting forces: a great stress on wage relativities and other conditions as a basis for making claims against the employer when one's group is held to have 'fallen behind'; unions outbidding each other as they seek to make changes in the norms regulating their members' wages and working conditions; and employers bidding up earnings over agreed rates as they compete for scarce labour. The whole adds up to the familiar leap-frogging image, with the impression that the game is being played at a faster and faster pace. Disorder breeds disorder.

Problems of Integration

What Flanders and Fox dwell upon in their proposals for the reconstruction of the normative order, is what we would term approaches to system integration. The central issue for them is 'whether the whole normative framework governing the production and distribution of wealth becomes further fragmented and splintered in a manner which threatens further disorder, or whether we are still capable of reconstructing larger areas of agreement upon which larger units of regulation can rest'.[1] The solution, if it is to be found, they see as multi-faceted and must be implemented at different but interrelated levels of economic life. The enterprise is viewed as an important source of normative integration for the diversity of work groups which it encompasses, and, to that end, comprehensive productivity agreements and job evaluation schemes are commended. Industry-wide regulation is seen as complementary to this, with the employers' associations and relevant trade unions seeking to provide guide lines for action:

> The aim here could be the long term pursuit of normative agreement within the industry on ways of measuring and rewarding different kinds of work, on methods of relating changes in rewards to changes in productive techniques, on the criteria which companies should apply when concerned with general wage or salary increases, on standards of labour utiliza-

[1] *British Journal of Industrial Relations*, VII, 1969, p. 174.

tion and definitions of work roles, on career structures and promotion criteria, on the handling of disciplinary and redundancy issues, and other matters of normative contention.[1]

There remains the question of inter-industry normative regulation. The potential role of public bodies like the National Board for Prices and Incomes, or the Commission on Industrial Relations, is here indicated. Such agencies would seek to find or suggest areas of normative agreement and attempt to establish regulative norms: reviewing and encouraging the development of procedural norms and, where necessary, testing claims of a substantive kind. The public scrutiny involved would serve not only to articulate and, if necessary, clarify the basis of normative regulation, but could also be regarded as a countervailing pressure against the processes of fragmentation.

In turning now to comment on this analysis it is first of all necessary to recognize that the diagnosis and solution proffered is in terms of promoting a pluralist society. Since this is the context in which the whole argument is placed, one could wish for a fuller delineation of the pluralist model. However, reference is made to the existence of a wide variety of relatively autonomous but interacting groups and agencies which create their own norms. They are, so to speak, expressions of voluntary action and serve to sustain and encourage freedom of contract and association, ideas and ideologies. The model is contrasted on the one side with a coercive authoritarian state in which order is imposed and conflicts suppressed, and, on the other side, with anarchy—in which social order is shattered by the excessive fragmentation of normative regulation discussed earlier. Since Fox and Flanders allow that a pluralist society has some degree of disorder within it, then a movement towards anarchy is a built-in risk. Their thesis, however, is that if a society starts to move rapidly in that direction, then the authoritatian solution may be sought. Government attempts in the UK to restrict wage movements and intervene in collective bargaining, together with the continuing discussion of the role of law in creating order in the industrial relations sphere, are cited as indications of a drift towards the authoritarian solution. In their rejection of legal compulsion as a

[1] *British Journal of Industrial Relations*, VII, 1969, pp. 176–7.

means of inducing industrial relations order Fox and Flanders follow the view of the Donovan Commission.[1] The Commission maintained, for example, in commenting on 'unconstitutional actions' by employees, that 'as long as no effective method for the settlement of grievances exists no one can expect a threat of legal sanctions to restrain men from using the advantage they feel able to derive from sudden action in order to obtain a remedy for grievances which cannot be dealt with in an orderly fashion. Self-help has always been the response to the absence of 'law and order'. In industrial relations, 'law and order' can be created only by adequate collective bargaining arrangements'.[2]

But the question we may now raise is whether the controversy over the use of law in British industrial relations finally gets to the crux of the matter so far as the anomie of economic life is concerned. We have already seen in our discussion of Durkheim that an adequate social policy to respond to the problem of anomie had to concern itself with remedying inequality in the external conditions of the struggle. And it is precisely on this point that Goldthorpe has queried the solution which Fox and Flanders propound.[3] It is a solution (and this point we may note applies likewise to the Donovan Commission) which may grant a greater degree of formal rationality to the industrial relations sphere but will not provide stability because it does not of itself provide a moral basis to economic life. If there are marked inequalities of educational and occupational opportunity in a society, as is the case in the UK, then a Durkheimian perspective would suggest that these must be attacked if a social and industrial order which necessarily implies inequalities of condition is to achieve any moral stability. 'One need not assume that rank and file industrial employees resent the inferior life chances they have been accorded as keenly as the facts might warrant in order to claim that few will feel *morally* bound by the normative codes which govern their working lives.'[4] Accordingly, an incomes policy (whether framed in voluntary or statutory terms) might

---

[1] The Royal Commission on Trade Unions and Employers' Associations, 1965–68: Report Cmnd. 3623.
[2] op. cit., p. 136.
[3] See J. H. Goldthorpe, 'Social Inequality and Social Integration in Modern Britain', Presidential Address to the British Association for the Advancement of Science: Sociology Section, 1969.
[4] op. cit., pp. 13–14.

in reality, if the wider structural inequalities continue to operate, have an effect opposite to the intended effect of inducing stability. It might serve in practice to widen the reference groups that are utilized when pay claims are made:

> . . . what are sometimes called the 'educative' functions of incomes policy may well have the effect of undermining the viability of such policy. To the extent that evaluations of income and other economic differences do become less confused and obscure, there is little reason to suppose that what will emerge will be greater consensus from one group or stratum to another: the far more likely outcome, given the prevailing degree of inequality, is that conflicts will become more clearly defined and more widely recognized—that the anomic state of economic life will be made increasingly manifest.[1]

The difference between Fox and Flanders' position on the one hand, and Goldthorpe's on the other, is certainly of some importance in terms of policy implications. Both positions diagnose a state of anomie in the economic sphere and both are clearly opposed to authoritarian solutions. But the implication of Goldthorpe's argument is that Fox and Flanders offer a faulty diagnosis, and hence a suspect cure, because they ignore the wider structural aspects which, while in some senses external to economic life, necessarily impinge upon it. Further, a policy of rationalizing the economic order in procedural and substantive terms might exacerbate rather than ameliorate the problems. And this might, indeed, in its turn lead to a move in the direction of an authoritarian solution—the very thing Fox and Flanders wished to avoid. Goldthorpe's own approach to a solution is more radical, namely— that to counteract anomic tendencies in economic life, the entire structure of power and advantage in society must become more capable of rational and moral justification. It must become a more 'principled', and hence a more widely acceptable and legitimate, social order. However, he is not sanguine about the likelihood of such a development in terms of the political realities of the British situation, and predicts rather the persistence of 'marked inequality and thus of chronic unrest and of general economic infighting between interest groups under the rules

[1] op. cit., p. 18.

mainly of "catch as catch can". Such a forecast is indicated by the fact that the egalitarian restructuring of our society, which could only be achieved as a work of political will, expertise and force, does not appear to be on the agenda of any major political party.'[1]

(5) *The Merton Anomie Paradigm: Applications and Critiques*

One of the most notable and influential contributions to the study of anomic behaviour has obviously been that of Robert Merton. His elaboration of the anomie theme in his seminal essay *Social Structure and Anomie*[2] derived from his interest in explaining deviant behaviour:

> Our primary aim is to discover how some social structures exert a definite pressure upon certain persons in the society to engage in non-conforming rather than conforming conduct.[3]

Merton thereupon proposes an analytical distinction between cultural goals and institutional norms. Prevailing goals are seen as providing a frame of reference for the aspirations of individuals in a particular society, and the institutional norms refer to regulatory practices through which goals may legitimately be realized in that society. The empirical relationships between cultural goals and institutional norms are however very complex. Where a rough balance exists between them, Merton suggests that such societies may be described as integrated, even though social changes may be observed. This situation he distinguishes in ideal-type terms from two forms of mal-integrated society: those which are stable but inflexible traditional societies with a ritualistic

---

[1] op. cit., p. 25.
[2] In *Social Theory and Social Structure*, Free Press, 1957, pp. 131–60.
[3] ibid., p. 132.

emphasis on conforming to institutional practices, and those whose members stress the importance of achieving certain goals without placing much emphasis on what institutional means are to be regarded as legitimate. The first of these is not too dissimilar in conception to Durkheim's use of mechanical solidarity which, in a situation of social change and development, is regarded as pathological. The second reflects a Durkheimian kind of interest in societies which have inadequate regulatory laws. That these norms have a moral character is implicit in Merton's distinction between technical norms for achieving ends, and institutional norms which may proscribe certain technical norms no matter how efficacious they may be. It is this second type of society that Merton focuses upon, and he chooses the United States as a case in point arguing that 'the process whereby exaltation of the end generates a literal *demoralization*, i.e. a de-institutionalization of the means occurs in many groups where the two components of the social structure are not highly integrated'.[1]

In his essay, Merton takes the success theme in American culture to illustrate his contention—success in making money being the operative criterion. The choice of this particular cultural goal brings us centrally into the sphere of economic life again. Although Merton claims he could have chosen other cultural goals for illustrative purposes, the actual choice has strong

*Merton's Typology of Modes of Individual Adaptation**

	Modes of Adaptation	Cultural Goals	Institutionalized Means
I	Conformity	+	+
II	Innovation	+	—
III	Ritualism	—	+
IV	Retreatism	—	—
V	Rebellion	±	±

+ = acceptance
— = rejection
± = rejection of prevailing values and substitution of new values

*Source: Merton, p. 140.

[1] *Social Theory and Social Structure*, Free Press, 1957, p. 136.

THE MERTON ANOMIE PARADIGM

echoes of Durkheim in it, since it is clearly recognized that the goal of monetary success is one which one is always striving towards rather than achieving in absolute terms. Now because Merton is interested in explaining deviant behaviour he goes on to consider what kind of responses individuals may make if they are unable or unwilling to achieve this cultural goal. This leads him to present his by now familiar paradigm (shown opposite).

With reference to the success goal, deviant responses II to V may be briefly indicated as follows:

Innovation: 'Despite our persisting open class ideology, advance towards the success goal is relatively rare and notably difficult for those armed with little formal education and few economic resources. The dominant pressure leads towards the gradual attenuation of legitimate, but by and large, ineffectual, strivings and the increasing use of illegitimate, but more or less effective, expedience.'[1]

Ritualism: 'It involves the abandoning or scaling down of the lofty cultural goals of great pecuniary success and rapid social mobility to the point where one's aspirations can be satisfied. But no one rejects the cultural obligation to attempt "to get ahead in the world", though one draws in one's horizons one continues to abide almost compulsively by institutional norms.'[2]

Retreatism: the individual who abandons formerly esteemed cultural goals and institutional norms once regarded as relevant guides to his behaviour is described by Merton as the true alien. The individual is specialized, socially disinherited.

Rebellion: 'In our society, organized movements for rebellion apparently aimed to introduce a social structure at which the cultural standards of success would be sharply modified and provision would be made for a closer correspondence between merit, effort and reward.'[3]

Broadly it might be argued that Merton's paradigm has a sensitizing function which may be of help in seeking to understand the processes of social life. It should be recognized that the paradigm itself is incomplete. In addition to the five modes of adaptation which he spells out, the three categories of acceptance (+) rejection (—) and rejection of prevailing values and sub-

---

[1] *Social Theory and Social Structure*, Free Press, 1957, pp. 145-6.
[2] ibid., pp. 149-50.   [3] ibid., p. 155.

stitution of new values ($\pm$), to be logically exhaustive, include the following:

Mode of Adaptation	Cultural Goals	Institutionalized Means
W	$+$	$\pm$
X	$-$	$\pm$
Y	$\pm$	$-$
Z	$\pm$	$+$

This does draw attention to the fact that in the original categorization the most active category—of rejection of prevailing values and substituting new values—is somewhat played down. Whether, and in what circumstances, all of these logical categories have empirical referents might be usefully explored. However, for our purposes we make the point because the examples Merton uses, and his discussion, indicate some uneasiness as to whether they all fit neatly into his typology. For example, the retreatism of workers who are described as being in a state of 'psychic passivity' is really different in kind from the retreatism of, say, the retired worker who, through retirement, has suffered an abrupt break in established social relations.

The most useful empirical examples to which Merton draws our attention, as far as the study of industrial life is concerned, appear, we may suggest, to be the following:

1. *Innovatory behaviour on the part of businessmen*

Merton refers mainly to Sutherland's work on white collar criminality.[1]

Whether the behaviour is strictly unlawful may be a matter of judgment. A good example of innovatory behaviour which was condemned by Government committees in the UK but not treated as unlawful, relates to Bristol Siddeley Engines Ltd. The company was accused of having made exorbitant profits on its engines 'as a result of false and misleading representations on the part of the Company's estimating and negotiating committee'.[2]

[1] See for example E. M. Sutherland, 'White Collar Criminality' in *ASR* V, 1940, and 'Is "white collar crime" crime?' in *ASR*, X, 1945.
[2] Wilson Committee Report. HC 129, 1967–68, para. 83.

In the event the Company had to pay back to the Ministry of Technology, in 1967, £3,960,000. Even though the Third Special Report of the Committee for Public Accounts wrote of negligence and irresponsibility in the financial administration of the Company, and the Wilson Committee did not exonerate the Executive Director from foreknowledge of what was going on, legal action was not taken. On the general point, Sutherland maintains that even if one is dealing with what is literally white collar crime, prosecution will not always follow, partly because of the high status of the businessman. Furthermore, public resentment against such behaviour is relatively unorganized.

Other forms of innovatory behaviour are well illustrated in Dalton's study *Men who Manage*.[1] There he describes the various mechanisms by which managers did conjuring tricks with the accounts to give themselves greater freedom with the dispersal of materials and services. These might result in relatively simple activities such as fitting out a 'plush' office to put oneself one up in the status rivalry stakes with other managers; or in taking relatives and friends of associates on to the pay-roll for part-time employment. On the other hand, 'underground' actions with major implications could be undertaken. Dalton cites the case of Jessop, divisional manager in a chemical plant, who had been thwarted by his chief from implementing his ideas for changing a refinement process. Consequently he made a successful case for taking on twenty additional employees. They were never taken on (though fictitious incomes and roles were created for them). Because he had good relations with the Time and Auditing Departments he was able to use the money to buy new equipment secretly, and to experiment in a vacant building. When a little later his boss retired, the successful new process was triumphantly brought out into the open and labelled 'the Jessop process'. Jessop regarded the operation as small in scale to some known to him. He cited a major modernization programme which a former plant chief had instituted. The original appropriation of $30,000,000 was extended to $37,000,000, the extra $7,000,000 being used to promote empire building and so on, while formally the costs were being skilfully attributed to legitimate factors.[2]

In the examples we have cited the innovatory behaviour applies

[1] M. Dalton, *Men Who Manage*, Wiley, 1959.
[2] See Dalton, op. cit., pp. 32–3.

to the manipulation of means to achieve stated ends. The end may or may not be achieved in practice. Although Merton concentrates on innovatory behaviour among businessmen as far as industrial life is concerned, clearly it can be paralleled in other segments of industrial life. This is also brought out in Dalton's study. There he points out that theft may operate on an individual or collective basis at all ranks in an organization, and that in practice it is difficult sometimes to distinguish between 'theft' and 'unofficial reward'. So, for example, loosely controlled expense accounts may be offered to executives and justified in terms of the need 'to attract and hold top men'. More generally, he points to a common practice in industry of doing 'goverment jobs'—that is work performed during the firm's time and using the firm's equipment and materials as a 'favour' for works' foremen or managers as the case may be. So, for example, the local yacht club had many men among its membership from the Milo & Fruhling Companies. Dalton comments:

> Building additions to the club and maintenance of its plant, as well as of privately owned boats, drew on the stores and services of Milo & Fruhling. Repair work was charged to various orders. . . . Propeller shafts, buskings, fin keels, counterweights, pistons, head railings and the like were made and/or repaired for boat owners among the managers as well as their friends, in the community.[1]

These examples cited from Dalton in the main point to a diversion of resources away from what would appear to be immediate task requirements. Part of his aim was to demonstrate that the concepts of 'theft' and 'reward' derive their meaning from the social context. However, other forms of innovatory behaviour may be aimed at improving matters in the immediate task situation. This is the burden of Donald Roy's paper 'Efficiency and "the Fix"'.[2] There he describes the behaviour of machine operators who by collusion with inspectors, set up men, tool crib men, and stackmen, sought to by-pass formal shop routines relating to the borrowing of tools and equipment in order to facilitate production.

2. *The ritualistic behaviour of bureacrats*

Merton here has in mind the phenomenon of 'over conformity'

[1] See Dalton, op. cit., p. 205.　　[2] D. Roy, op. cit.

which he sees as a condition of acute status anxiety in a society which emphasizes the achievement motive. His main empirical reference is to the work of Blau[1] who contended that rigid adherence to procedure by bureaucrats was derived from a lack of security in important social relationships in the organization. Merton extends his comment in a further essay, 'Bureaucratic Structure and Personality:[2]

> Discipline readily interpreted as conformism with regulations, whatever the situation, is seen not as a measure designed for specific purposes but becomes an immediate value in the life-organization of the bureaucrat. This emphasis, resulting from the displacement of the original goals develops into rigidities and an inability to adjust readily. Formalism, even ritualism, ensues with an unchallenged insistence upon punctilious adherence to formalized procedures. This may be exaggerated to the point where primary concern with the conformity to the rules interferes with the achievement of the purposes of the organization in which case we have the familiar phenomenon of the technicism or red tape of the official'.[3]

Again it may be suggested, as in the case of innovation, that the distinction between individual behaviour and group-sanctioned behaviour is somewhat blurred in the discussion. For example, Blau's work (to which Merton refers) stresses the differences in competence of individuals in the same situation. It was the incompetent who could calm their own anxieties by the ritual of rule conformity and who strenuously resisted the institution of procedural changes. At the same time Merton draws attention to the *esprit de corps* which can develop among a group of bureaucratic functionaries such that ritualist adherence to red tape serves to operate as a defensive measure when the group's interests are threatened. Whether this latter form of behaviour should be defined as 'over-conformist' really depends on one's standpoint. What is implied here of course is that the form of ritualism is not properly defined as a neurotic response to change, but, as Crozier has argued, 'a very useful instrument in the struggle for power and control and in the protection of a group's area of action. Ritualism,

---

[1] P. M. Blau, *The Dynamics of Bureaucracy*, University of Chicago Press, 1955.
[2] In *Social Theory and Social Structure*, op. cit.   [3] *ibid.*, p. 199.

in such a context, if one considers the actors' frame of reference and not the whole organization, must be viewed as conformity to what is expected from him and no longer as over-conformity. It is the rational response and not a "professional deformation".[1]

It is probably fair to say that in Merton's discussion the ritualist bureaucrat (allowing for the ambiguities already alluded to) occupies a central place in the analysis of bureaucratic personality. Crozier suggests that this is to over-simplify. Other images of the bureaucrat may be located including innovating rebellious and submissive types. And one official may exhibit rebellious and ritualistic behaviour patterns. In discussing French provincial administration, for example, a picture is drawn of the prefect with discretionary powers (described by Crozier as an innovator, although, even given Merton's ambiguities, this is a little misleading, since there is nothing 'illegitimate' in principle about the actions he takes). He is backed up by submissive intermediary officials who identify with him. But the petty officials are described as ritualistic and *at the same time* rebellious:

> They are attached to the status quo and resent possible innovations as so many violations of the order which they must impose on the public. They feel themselves to be betrayed, and their position undermined by the prefect's initiative. Their strategy is a strategy of opposition and rebellion: they try to impose their ritualism on the prefect and to obtain some compensation for his trespassing on their jobs. We see here the importance of the gap created by centralization. Petty officials cannot make the necessary adjustments. Power to innovate is reserved for superior figures with prestige. As a result petty officials behave as extremely jealous ritualists for all practical purposes, and try to use to the utmost advantage the parcel of power involved in the rites imposed on them. At the same time they question the whole system and pose as rebels. Sentiments are more complex than one would expect from reading Merton. There is a sort of paradox in this respect: when a petty official obtains promotion to a middle rank that permits him to escape from the chicanery of petty regulations, he forgoes this theoretical rebellion and becomes humble and submissive.[2]

[1] M. Crozier, *The Bureaucratic Phenomenon*, Tavistock Publications, 1964, p. 199. [2] *ibid.*, p. 202.

We may here note, however, that by referring to theoretical rebellion Crozier is in fact implying a distinction between attitudes and behaviour. It is one thing after all to question a system, it is quite another to work actively for its overthrow.

3. *Retreatist behaviour among industrial workers*

Merton, in this case, alludes to Chinoy's well-known study *Automobile Workers and the American Dream*.[1] Chinoy was looking at a plant which, with a preponderance of unskilled jobs, offered very limited opportunities of getting into a managerial position (or even moving from an unskilled to a worker category.) Striving for success in the job sphere, among the majority of employees, was not common. There was a widespread lack of interest, hope, or desire even to become foremen (the acknowledged first rung on the promotion ladder) among the shop floor employees. Some nourished hopes of success in terms of setting up their own business or farm. A few escaped, some made abortive attempts, but the great majority eventually accepted the fact that the factory constituted the setting of their working life and scaled their aspirations down accordingly. At this point a more passive response to the work situation and one's life chances was noted.

What this signifies in terms of the life experience of the worker may vary somewhat. Chinoy points out that the disappearance of ambition does not always appear to result in a sense of disappointment or frustration. Some resign themselves contentedly to what they have achieved—perhaps a relative degree of security through seniority, or a relatively interesting or satisfying job compared to others in the plant. Others 'scarred by experience, resign themselves to a future in the factory without satisfaction but without resentment. They no longer demand much out of life except for some kind of job and some assurance that they can keep it.'[2] Others however find it more difficult to come to terms with what they clearly regard as defeat:

> ... if workers come to feel that they must stay in the factory because there is no opportunity in business or farming, if they do not have desirable jobs in the plant, if they began their careers with large ambitions and high hopes, or if they have

[1] E. Chinoy, *Automobile Workers and the American Dream*, Doubleday, 1955.
[2] op. cit., p. 122.

seen relatives or friends 'get ahead in the world', then their acceptance of a future in the factory is accompanied by bitterness and resentment aimed at themselves, at others or at the world in general.[1]

4. *The retreatism of the socially disconnected*

Here the focus is upon those who become apathetic as a result of social circumstances which place the individual in a social vacuum in which his life no longer seems to have any sense of direction or meaning. The ranks of the long-term unemployed provides many cases in point. E. Wight Bakke in his sensitive study *The Unemployed Man: A Social Study* reports that nearly all the unemployed workers he spoke to conveyed a sense of the feeling of being lost without work. As time goes on and they are still without work, the early confidence with which they engaged on the job search is sapped—to be replaced by apathy, sullenness, and a growing lack of self-respect, as they feel unable to discharge responsibilities towards their families. Bakke further observes:

> ... the sensitiveness to the importance of one's place as worker will cause him to get disheartened more quickly than the man who has less sensitiveness on this point. I was not surprised, therefore, to find among the men with whom I associated for several months that the most ambitious lost heart most quickly. The quality that *on the job* leads to rapid achievement of greater and greater satisfaction *off the job* leads to rapid retreat into hopelessness and discontent, despair and even sullenness. The incentive to work hard, the desire to push ahead, the ambition to perfect one's technique, these are basic qualities for satisfaction at work. They are just the qualities that make it hardest for a man to be out of work. ...
>
> With a job there is a future; without a job there is slow death of all that makes a man ambitious, industrious and glad to be alive.[2]

Other studies too have pointed to the way in which prolonged loss of a job tends to be associated with an overall withdrawal of social contacts—not only with formal groups such as trade unions,

---

[1] op. cit., p. 122.
[2] E. Wight Bakke, *The Unemployed Man*, Nisbet, 1933, p. 72.

clubs and churches, but informal visiting of friends, relatives, and so on.[1]

The studies we have cited were grounded in the depression years (between the wars) in which unemployment was summed up by Orwell as 'the frightful doom of a decent working man suddenly thrown on the streets after a lifetime of steady work, his agonized struggles against economic laws which he does not understand, the disintegration of families (and) the corroding sense of shame'.[2] It is however an experience still to be encountered: notably in regions of high unemployment which may be located in so-called full employment societies.[3] It is the unskilled workers living in areas which are heavily dependent upon basic industries, such as coal-mining, ship-building or steel who are particularly at risk. In the United States there is an ethnic dimension to this since Negroes, American Indians and Puerto Ricans, to take the more obvious examples are disproportionately represented among the ranks of the unskilled. Skill attainment of course is necessarily related to educational experience and opportunity, and in their carefully documented study Blau and Duncan show that the chances for Negroes to go beyond the grammar school level to more advanced education as compared with Whites have become relatively worse in recent decades. This, they argue, 'surely poses a serious challenge to the complacent belief that the position of the Negro is gradually improving particularly in democratic society that prides itself on the educational opportunities it offers to all its citizens'.[4] The structural sources of long term unemployment are not of course solely related to changes in the product market but also to technological change. In his evidence to the Senate Sub-Committee on Employment and Manpower in 1960, Charles Killingworth pointed out that employment of blue-collar workers in manufacturing had been falling off over the past few years whilst, at the same time, output was rising. Many of the hard core unemployed in depressed areas were those whose last job was in

---

[1] See for example M. Komarovsky, *The Unemployed Man and His Family*, Dryden 1940. B. Zawadski and P. Lazarsfeld, 'The Psychological Consequences of Unemployment' in *J. of Soc. Psy.*, VI (1935). E. Wight Bakke, *Citizens Without Work*, Yale University Press, 1940.
[2] G. Orwell, *The Road to Wigan Pier*, Penguin, 1962, p. 131.
[3] In, for example, A. Sinfield, *The Long Term Unemployed*, OECD, 1969.
[4] P. Blau and O. Duncan, *The American Occupational Structure*, Wiley, 1967, p. 227.

manufacturing.[1] Even if one accepts the view that technological advance broadly improves the chances of the population in general for upward social mobility, there can be no doubt that those same advances worsen the chances for some and seriously disrupt occupational careers. In a more recent paper Killingworth has written of rising unemployment in periods of record prosperity as 'the most ominous economic problem in the United States today'.[2]

His thesis is that 'the rising unemployment of the past decade has been caused primarily by the interaction between new technology and the changing consumption patterns of a mature mass-consumption society. This interaction has caused sharp employment declines in some sectors and has helped to create labour shortages in still other sectors. The overall effect has been growing imbalance in the labour market with a great surplus of unskilled, poorly-educated workers co-existing with serious shortages of many kinds of highly-educated workers.'[3] What we are suggesting therefore is that the retreatist mode of adaptation cannot be relegated to the status of historical relic in so far as long term unemployment is still part and parcel of even advanced industrial societies for large numbers of people. But one should not, of course, infer that all unemployed persons necessarily adapt in this way. Nor should one ignore the fact that unemployment, when recognized as a social problem, can promote social tension. Some unemployed workers adapt in the way we have indicated, but others show a willingness to support mass movements. Such movements may indeed not only help the individual to overcome personal feelings of anxiety and futility, but also serve to question the credibility of the ruling politicians. As Kornhauser has pointed out 'the inability to terminate such crises, or at least to mitigate them, makes political and economic rulers highly vulnerable to direct pressures, such as unemployment demonstrations, marches on legislators and political violence in the streets'.[4]

[1] See Charles C. Killingworth, 'Implications of Automation for Employment and Manpower Planning' in S. Marcson (ed.), *Automation, Alienation and Anomie*, Harper, 1970).
[2] Charles C. Killingworth, 'Structural Employment in the United States' in J. Stieber (ed.), *Employment Problems of Automation and Advanced Technology*, Macmillan, 1966.
[3] op. cit., pp. 128-9.
[4] W. Kornhauser, *The Politics of Mass Society*, Routledge & Kegan Paul, 1960, p. 167.

5. Rebellion

Merton does not apply the category of rebellion to economic life but there are one or two points which are worth developing. The general distinction is made between retreatism as a privatized mode of adaptation as opposed to rebellion which implies a collectivist mode of adaptation which explicitly rejects and in some way challenges the existing order. Such rebellion may be confined to relatively powerless elements in a community who in fact develop and live within sub-cultures of their own. An occupation may form the basis of such a sub-culture. A good example of this is Becker's essay on the dance musicians.[1] Becker described one group of jazzmen who were extremists in that they totally rejected American society:

> Every interest of this group emphasized their isolation from the standards and interests of conventional society. They associated almost exclusively with other musicians and girls who sang or danced in night clubs in the North Clark Street area of Chicago and had little or no contact with the conventional world. They were described politically thus: 'They hate this form of government anyway and think it's real bad.' They were unremittingly critical of business and labour, disillusioned with the economic structures and cynical about the political process and contemporary political parties. Religion and marriage were rejected completely, as were American popular and serious culture, and their reading was confined solely to the more esoteric *avant garde* writers and philosophers. In every case they were quick to point out that their interests were not those of the conventional society and that they were thereby differentiated from it.[2]

In this example, however, the categorization of rebellion describes a group which has been able to adopt an alternative way of life to that of the conventional 'squares' without actually attempting to wrest power from them. It may display its contempt for the conventional world and its values by refusing to assimilate, but as a group it accommodates to that on-going world and does not seek to transform it by group action. Compare this with groups of

---

[1] Howard S. Becker, *Outsiders. Studies in the Sociology of Deviance*, chapter V. Free Press, 1963.
[2] op. cit., p. 98.

industrial workers who actively see themselves in opposition to the wider society. For them the strike, or other weapons at their disposal, has the character of (at least) a small scale revolt against society. Writing of the high propensity of certain occupational groups to strike, as reflected in international strike statistics based upon numbers of working days lost, Kerr and Siegel suggest that many of them have the character of an 'isolated mass':

> The miners, the sailors, the longshoremen, the loggers and to a much lesser extent, the textile workers, form isolated masses, almost a 'race apart'. They live in their own separate communities: the coal patch, the ship, the waterfront district, the logging camp, the textile town. These communities have their own codes, myths, heroes and social standards. There are few neutrals in them to mediate conflict and dilute the mass. All people have grievances but what is important is that all the members of each of these groups have the same grievances.... The employees form a largely homogeneous, and undifferentiated mass—they all do about the same work and have about the same experiences.[1]

The sense of a mass grievance against society as a whole, or its ruling interests, helps to explain why the decision to strike and to stop striking may not be understood as an economic calculation of costs set against benefits.

Merton argues that for rebellion to pass into organized political action the groups which challenge the existing social structure must be 'possessed of a new myth'. We may recall Sorel's myth of the General Strike as a case in point. Of such myths, Sorel maintained that they are not descriptions of things, but expressions of a determinism to act.[2] They are, in short, expressions of group convictions in the language of movement. The image and even the experience of strike action may not match actual accomplishments but its existence can kindle action. Take a more localized but long and bitter recent strike dispute at Pilkington's Glass Works in St Helen's, Lancashire. There we find a striker speaking of a moment of solidarity:

---

[1] C. Kerr and A. Siegel, 'The Interindustry Propensity to Strike' in C. Kerr, *Labor and Management in Industrial Society*, Doubleday, 1964, pp. 109-10.
[2] See G. Sorel, *Reflections on Violence*, Collier Books, 1961.

It was fantastic that afternoon. We could have done anything. We could have stopped the world. We didn't give a monkey's for the rain, the Bobbies, Pilkington's, the union. It was . . . bloody great.[1]

It remains for us to say something about the significance of the conformity category in Merton's typology. Although Merton's treatment of anomie was a conscious attempt to develop Durkheim's analysis, recently Horton has questioned Merton's treatment of anomie arguing that his position in one important respect is different from Durkheim's:

For Durkheim, anomie was endemic in such (success oriented) societies not only because of inequality in the conditions of competition, but, more importantly, because self-interested striving (the status and success goals) had been raised to social ends. The institutionalization of self-interest meant the legitimization of anarchy and amorality. To maximise opportunities for achieving success would in no way end anomie. Durkheim questioned the very values which Merton holds constant.[2]

What Horton is in fact suggesting is the anomic character of the conformist response in a success-oriented culture. However, this may be a misinterpretation of Merton's position since the essence of the analysis is on what happens in a society in which there are strains towards anomie. Merton makes the general point, we may recall, that 'any cultural goals which receive extreme and only negligibly qualified emphasis in the culture of a group will serve to attenuate the emphasis on institutionalized practices and make for anomie'.[3] It is the social order itself which is labelled anomic. In a situation of acute anomie Merton sees the system of institutional controls as having broken down, and calculations of personal advantage and fear of punishment appear to be the only regulatory agencies. And he seems to be suggesting that the stress on the success goal is particularly likely to promote anomie, not only because of the differential placing of social groups in relation to

---

[1] C. Barker, 'The Pilkington Strike' in I.S., 1970, p. 5.
[2] J. Horton, 'The De-humanization of Anomie and Alienation: a Problem in the Ideology of Sociology' in *BJS.*, XV, No. 4, pp. 294–5.
[3] 'Social Structure and Anomie: Continuities', in Merton, op. cit., p. 181.

the opportunity structure, *but also* because success itself is a never-attainable goal even for the conformer. Such a society, we may say, has an in-built tendency to get overheated. In this respect Merton's position is not nearly so conservative as Horton would have us think. Matters are perhaps somewhat clarified by his later essay *Anomie, Anomia, and Social Interaction: Contexts of Deviant Behaviour*.[1] The distinction between anomie and anomia is more carefully worked out than in the original essays. Anomie refers to a property of a social system:

> It refers to a breakdown of social standards governing behaviour and so also signifies little social cohesion. When a high degree of anomie has set in, the rules once governing behaviour have lost their savour and their force. Above all else they are deprived of legitimacy ... there is no longer a widely shared sense within the social system, large or small, of what goes or does not go, of what is justly allowed by way of behaviour and what is justly prohibited, of what may be legitimately allowed in the course of social interaction.[2]

Anomia, following Srole,[3] is used to designate a state of mind of the individual in an anomic situation. Further, the important distinction between the anomia of success and the anomia of deprivation is re-emphasized. Innovation, ritualism, retreatism, rebellion are possible responses to the anomie of deprivation. Of course, what is implied here is that there has to be a sense of deprivation in relation to a cultural goal for the subjective state of anomia to be present (no matter how a person stands objectively in relation to the opportunity structure through which the goal may be approximated). The anomie of success is, as we have already seen, the response which particularly fascinated Durkheim. And to his remarks about the way men may be driven on by insatiable desires, such that a goal once achieved does not provide lasting satisfaction, Merton adds that there may well be social pressures which promote this response:

---

[1] In M. B. Clinard (ed.), *Anomie and Deviant Behaviour*, Free Press, 1964, pp. 213–42.
[2] *ibid.*, p. 226.
[3] L. Srole, Social Integration and Certain Corollaries: An Exploratory Study, *ASR*, Vol. 21, 1956.

Social pressures do not easily permit those who have climbed the rugged mountains of success to remain content; there is no rest for the weary. . . . More and more is expected of these men by others and this creates its own measure of stress. Less often than one might believe, is there room for repose at the top.[1]

Where there is perhaps an omission in Merton's approach as compared with Durkheim's is that there appears to be nothing equivalent in his theory to the part played by the forced division of labour in Durkheim. To this extent, the possibilities of one group exploiting another as a result of power disparities between them does not receive independent attention. But the issues raised by Durkheim in his treatment of the forced division of labour do, of course, find a central place in Marx's work. This we choose to make a starting-point in Part III where we discuss the alienation theme with particular reference to the sphere of work.

[1] L. Srole, Social Integration and Certain Corollaries: An Exploratory Study, *ASR*, Vol. 21, 1956, p. 221.

Part Three

*Individual and Society:
Alienation—and a New Humanity?*

Introduction

To suggest that a person or group is in a state of alienation is to invite the following kinds of questions:

(*a*) From what are they alienated?
(*b*) How has this come about?
(*c*) What are the marks or circumstances by which we judge them to be alienated?

And since there is usually the strong implication that the state of alienation is undesirable, these questions are commonly followed by reflections on whether the condition is, in principle, curable and, if so, how. People who ask questions of this order—and theologians, philosophers, social scientists and psychiatrists often do—may be said to share a common perspective—but the commonality rests in the questions that are asked and certainly not in the answers that are given.

In this chapter we will indicate some of the ways in which the alienation theme has been handled in sociological analysis, and particularly with reference to industrial societies.

(*1*) *Work and Alienation in Marx*

It is to the *Economic and Philosophical Manuscripts* of 1841[1] that one first properly turns in approaching Marx's analysis of alienation in industrial society. In the chapter on Estranged

[1] See K. Marx, *Economic and Philosophical Manuscripts of 1844*, edited and with an introduction by D. L. K. Struik, Lawrence & Wishart, 1970.

Labour he asks what is the relationship between the worker and production. He propounds two themes and deduces a third from them:

1. The product of labour exercises power over the worker—it becomes an alien object independent of the producer.

> It is true that labour produces for the rich wonderful things—but for the worker it produces privation. It produces palaces—but for the worker hovels. It produces beauty—but for the worker, deformity. It replaces labour by machines, but it throws a section of the workers back to a barbarous type of labour, and it turns the other workers into machines. It produces intelligence—but for the worker stupidity, cretinism.[1]

2. The activity of work itself is an alienating activity and as such promotes self-estrangement and is expressed in the emasculation of the worker's physical and mental energy and his personal life.

> What, then, constitutes the alienation of labour? First, the fact that labour is *external* to the worker, i.e. it does not belong to his essential being; that in his work, therefore, he does not affirm himself but denies himself, does not feel content but unhappy, does not develop freely his physical and mental energy but mortifies his body and ruins his mind. The worker therefore only feels himself outside his work, and in his work feels outside himself. He is at home when he is not working, and when he is working he is not at home. His labour is therefore not voluntary, but coerced; it is *forced labour*. It is therefore not the satisfaction of a need; it is merely a *means* to satisfy needs external to it. Its alien character emerges clearly in the fact that as soon as no physical or other compulsion exists, labour is shunned like the plague. External labour, labour in which man alienates his self, is a labour of self-sacrifice, of mortification. Lastly, the external character of labour for the worker appears in the fact that it is not his own, but someone else's, that it does not belong to him, that in it he belongs, not to himself, but to another. Just as in religion the spontaneous

---

[1] K. Marx, *Economic and Philosophical Manuscripts of 1844*, p. 110.

activity of the human imagination, of the human brain and the human heart, operates independently of the individual—that is, operates on him as an alien, divine or diabolical activity—so is the worker's activity not his spontaneous activity. It belongs to another; it is the loss of his self.[1]

3. Because the worker is estranged from the things which he has created and from himself, the essential nature of man is violated.

... Estranged labour estranges the *species* from man. It changes for him the *life of the species* into a means of individual life. First it estranges the life of the species and individual life, and secondly it makes individual life in its abstract form the purpose of the life of the species, likewise in its abstract and estranged form.

Indeed, labour, *life activity*, *productive life* itself, appears in the first place merely as a *means* of satisfying a need—the need to maintain physical existence. Yet the productive life is the life of the species. It is life engendering—life. The whole character of a species—its species character—contained in the character of its life activity; and free, conscious activity is man's species character. Life itself appears only as a *means to life*.[2]

So then, that which distinguishes the human animal from other animals, his consciousness, the very source of his freedom, is distorted by the fact of estranged labour so that his essential being becomes a mere means to his existence. From this Marx derives the further implication:

The proposition that man's species nature is estranged from him means that one man is estranged from the other, as each of them is from man's essential nature.[3]

On the basis of this analysis Marx concludes that both wages and private property must be regarded as *consequences* rather than *causes* of alienation in the first instance. On the wages front Marx

[1] K. Marx, *Economic and Philosophical Manuscripts of 1844*, pp. 110–11.
[2] *ibid.*, pp. 112–13.   [3] *ibid.*, p. 114.

argues that to increase wages would not release the worker from the bondage of alienation. He would simply be a better paid slave. And movements to promote egalitarian wages systems are not an adequate response either: it is simply to conceive of society as an abstract capitalist. Rather his position is this: 'Wages are a direct consequence of estranged labour, and estranged labour is the direct cause of private property. The downfall of the one must be the downfall of the other.'[1] This helps us to appreciate more fully his later comments in the *Economic Philosophical Manuscripts* on the role of money and the division of labour respectively.

> By possessing the *property* of buying everything, by possessing the property of appropriating all objects, *money* is thus the *object* of eminent possession. The universality of its *property* is the omnipotence of its being. It therefore functions as almighty being. Money is the *pimp* between man's need and the object, between his life and his means of life. But that *which* mediates *my* life for me, also *mediates* the existence of other people for *me*. For me it is the *other* person.[2]

Certainly money, because it is a medium of exchange, binds men to society and to one another, but because it is the bond of all bonds it is also the universal agent of separation. It reflects, expresses and maintains the alienated ability of mankind:

> Money, then, appears as this *overturning* power both against the individual and against the bonds of society, etc., which claim to be *essences* in themselves. It transforms fidelity into infidelity, love into hate, hate into love, virtue into vice, vice into virtue, servant into master, master into servant, idiocy into intelligence and intelligence into idiocy.[3]

As for the division of labour, Marx reminds his readers of the views of some of the leading political economists: Adam Smith, Say, J. S. Mill. Together with the use of machinery it is seen as promoting wealth, and is indeed the basis of large-scale production in factories. Characteristically, each worker is permitted as small

---

[1] K. Marx, *Economic and Philosophical Manuscripts of 1844*, p. 118.
[2] *ibid.*, pp. 165–66.   [3] *ibid.*, p. 169.

a sphere of work as possible, notwithstanding the fact that he may be more intelligent or capable than could be demonstrated through such activity. But since Marx has already depicted labour as alienated activity, the division of labour is seen as 'the economic expression of the *social character of labour* within the estrangement'.[1]

Marx sketches (it is no more than that) the changing character of private property from that based on land ownership to that based on industrial capital. The victory of industrial capital with its accompanying factory system leads him to declare that 'it is only at this point that private property can complete its dominion over man and become, in its most general form, a world-historical power'.[2] It is, in short, the most developed expression of treating men as a commodity, and labour as a cost to capital.

> The *value* of the worker as capital rises according to demand and supply, and even *physically* his *existence*, his *life*, was and is looked upon as a supply of a *commodity* like any other. The worker produces capital, capital produces him—hence he produces himself, and man as *worker*, as a *commodity*, is the product of this entire cycle. To the man who is nothing more than a *worker*—and to him as a worker—his human qualities only exist insofar as they exist for capital *alien* to him. Because man and capital are foreign to each other, however, and thus stand in an indifferent, external and accidental relationship to each other, it is inevitable that this foreignness should also appear as something *real*. . . . The existence of capital is *his* existence, his *life*; as it determines the tenor of his life in a manner indifferent to him.[3]

It is communism which is portrayed as the solution to the alienation of man and his work and this is poetically expressed in the Hegelian language of the *Economical and Philosophical Manuscripts*:

> *Communism* as the *positive* transcendence of private property, as *human self estrangement*, and therefore as the real *appropria-*

---

[1] K. Marx, *Economic and Philosophical Manuscripts of 1844*, p. 159.
[2] *ibid.*, p. 131.   [3] *ibid.*, p. 120.

tion of the human essence by and for man; communism therefore as the complete return of man to himself as a *social* (i.e. human) being—a return become conscious, and accomplished within the entire wealth of previous development. This communism, as fully developed naturalism, equals humanism, and as fully developed humanism equals naturalism; it is the *genuine* resolution of the conflict between man and nature and between man and man—the true resolution the strife between existence and essence, between objectification and self-confirmation, between freedom and necessity, between the individual and the species. Communism is the riddle of history solved, and it knows itself to be this solution.[1]

Although this is the desired end, Marx is prepared to acknowledge that between the development of revolutionary consciousness—the idea of the need to abolish private property and the realization of the idea in action, one will be involved in 'a very severe and protracted process'.[2] But when realized it is nothing less than the discovery of man as man in a real society:

The *human* essence of nature first exists only for *social* man; for only here does nature exist for him as a *bond* with *man*—as his existence for the other and the others existence for him—as the life element of human reality. Only here does nature exist as the *foundation* of his own *human* existence. Only here as what is to him his *natural* existence become his *human* existence, and nature become man for him. Thus *society* is the unity of being of man with nature—the true resurrection of nature—the naturalism of man and the humanism of nature both brought to fulfillment.[3]

As man's essential being is so disclosed he is truly, and as a matter of enduring reality, to be reckoned a rich man profoundly endowed with all the senses. A glimpse of this future bliss is suggested to Marx by what he had observed of French socialist workers. 'Company, association and conversation, which again has society as its end, are enough for them; the brotherhood of man is no mere phrase for them, but a fact of life, and the nobility of man shines upon us from their work-hardened bodies.'[4]

[1] K. Marx, *Economic and Philosophical Manuscripts of 1844*, p. 135.
[2] *ibid.*, p. 154.   [3] *ibid.*, p. 137.   [4] *ibid.*, p. 159.

But if the things which get in the way of a man's social existence are alienating, then religion, family, state, law, morality, science, art are to be so understood. Marx maintains all such forms of estrangement will be ended since the appropriation of *human* life is involved in the positive transcendence of private property. Further, although it is the working class which is seen as the agent of emancipation, it is the emancipation of all—worker and non-worker—that is to be realized, since in a relationship of servitude human reality is distorted even for the dominator. Indeed the morality which political economy assumed could hold capitalist and worker in thraldom despite the power which the capitalist possessed over the lives of others. The ascetic demands of 'this science of marvellous industry' imply a rejection of life itself:

> ... its true ideal is the *ascetic* but *extortionate* miser and the *ascetic* but *productive* slave. Its moral ideal is the *worker* who takes part of his wages to the savings bank. ... Thus political economy—despite its worldly and wanton appearance—is a true moral science, the most moral of all the sciences. Self-renunciation, denunciation of life and all human needs, is its principal thesis. The less you eat, drink and buy books; the less you go to the theatre, the dance hall, the public-house; the less you think, love, theorize, sing, paint, fence, etc., the more you *save*—the *greater* becomes your treasure which neither moths nor dust will devour—your *capital*. The less you *are*, the less you express your own life, the greater is your *alienated* life, the more you *have*, the greater is the store of your estranged beings. Everything which the political economist takes from you in life and in humanity, he replaces for you in *money* and in *wealth*; and all the things which you cannot do, your money can do. It can eat and drink, go to the dance hall and the theatre; it can travel, it can appropriate art, learning, the treasures of the past, political power—all this it *can* appropriate for you—it can buy all this for you: it is the true *endowment*. Yet being all this it is *inclined* to do nothing but create itself, buy itself; for everything else is after all its servant. And when I have the master I have the servant and do not need his servant. All passions and all activity must therefore be submerged in *greed*. The worker may only have enough

for him to want to live, and they only want to live in order to have that.[1]

We have now sufficiently indicated that what is implied in Marx's treatment of the alienation concept is an image of man's essential nature. A view of social reality and human freedom is of course written in. *The Economic and Philosophical Manuscripts* are indeed the site of a collision between Hegelian philosophy and classical political economy. The significance of the collision has been well expressed by M. Nicolaus:

> This is a battle of methods, of ways of seeing and explaining the world, a struggle between disparate epistemologies. Here the dialectic power of German idealism struggles like Hercules against the giant, Antaeus, the son of Earth; and, it must be said, the outcome is the same as in that mythical trial: philosophy lifts its antagonist off the ground, away from the source of his strength and crushes him in mid-air. Thus Marx seizes upon the capitalist production process, its relations of property, together with its system of exchange and circulation, and lifts this entire edifice of empirical fact and empirical fancy into the Hegelian air, where he compresses the pragmatic giant into the single concept of 'alienated labour'. And Marx aims higher than Hercules; he not only crushes his antagonist, but he also believes that he can then reconstitute him on a higher level by unfolding the content of the fundamental core to which he has been reduced.[2]

It is, in short, a victory of metaphysics over empiricism. Does one therefore dismiss the alienation concept as a romantic element in the young Marx which he subsequently abandoned? The romanticism, if such it be, is certainly not confined to *The Economic and Philosophical Manuscripts*. In *The German Ideology* (1846), for example, are located his celebrated comments on communism and the abolition of the division of labour. Labour is not a voluntary activity but rather an expression of the fact that his own

---

[1] K. Marx, *Economic and Philosophical Manuscripts of 1844*, p. 150.
[2] Martin Nicolaus, *Proletariat and Middle Class in Marx: Hegelian Choreography and the Capitalist Dialectic*, Studies on the Left, Jan. 1967, pp. 26–7.

work is an alien power opposed to him. Instead of controlling it, he is enslaved by it:

> For as soon as the distribution of labour comes into being, each man has a particular, exclusive sphere of activity, which is forced upon him from which he cannot escape. He is a hunter, a fisherman, a shepherd, or a critical critic, and must remain so if he does not want to lose his means of livelihood; while in a communist society, where nobody has one exclusive sphere of activity but each can become accomplished in any branch he wishes, society regulates the general production and thus makes it possible for me to do one thing today and another tomorrow, to hunt in the morning, fish in the afternoon, rear cattle in the evening, criticize after dinner, just as I have a mind, without ever becoming hunter, fisherman, shepherd or critic.[1]

One may fairly point out that there is clearly a utopian strain in such statement. One may ask in practical terms where the rifles, rods and books are to come from to maintain this idyllic existence and also what precisely is involved in the phrase 'society regulates the general production and makes it possible for me to do one thing today and another tomorrow'. But whatever it means in a positive sense it certainly implies the over-throw of the state since it is the expression in Marx's view of the domination of the ruling classes (with whom are consolidated the ownership of property).

At the same time, Marx does proceed in *The German Ideology* to give a fairly extensive empirical analysis of the development of the division of labour up to the capitalist societies of his own day. This includes a basic distinction between the division of labour based on towns and that based upon agriculture. In both cases however work is regarded as forced activity 'a subjection which makes one man into a restricted town animal and the other into a restricted country animal'.[2] And the clear connection is drawn between the existence of private property and power over individuals who labour. In the town itself the following distinctions emerge so far as the division of labour is concerned:

---

[1] K. Marx and F. Engels, *The German Ideology*, Lawrence & Wishart, 1965, pp. 44–5.
[2] *ibid.*, p. 65.

(a) *Guild labour and non-Guild labour.* The apprentices and journeymen in particular crafts were bound to one another, but under the patriarchal rule of the Guild master. They typically aspired to become Guild masters themselves and therefore were interested in maintaining, rather than revolting against, the existing social order. The unskilled labourers, while having no such inhibitions, could only revolt in a sporadic and unorganized way and were in practical terms powerless to alter their conditions.

We may observe here that it is frequently argued that the craftsman may be treated as the archetypical non-alienated worker. Marx's position is somewhat different from this. In the mediaeval guild system, given the limited areas of commerce, the division of labour was not highly developed, so that: 'every man who wished to become a guild master had to be proficient in the whole of his craft. Thus there is found with the mediaeval craftsman an interest in their special work and in proficiency in it, which was capable of rising to a narrow artistic sense. For this very reason, however, every mediaeval craftsman was completely absorbed in his work, to which he had a contented, slavish relationship, and to which he was subjected to a far greater extent than the modern worker, whose work is a matter of indifference to him.'[1] It is the phrase 'contented, slavish relationship' which should especially be borne in mind. It is coercion which is the critical consideration, not feelings of dissatisfaction.

(b) *The separation of production from commerce.* The extension of the market which this implies and the concomitant increased contact between the towns made further specialism possible. This structural change Marx sees as transforming the individual burghers of individual towns into a *class*—conscious of its common interests with burghers elsewhere *vis-à-vis* other classes—while, at the same time, they may be in a condition of competition with each other. And the same change also provides the impetus for the growth of manufacturing industry, notably textiles, which superseded guild based industry.

(c) *The international division of labour.* This is the end consequence of (b) above. It is the basis of universal competition. 'It produced world history for the first time, insofar as it made all civilized nations and every individual member of them dependent

---

[1] K. Marx and F. Engels, *The German Ideology*, Lawrence & Wishart, 1965, p. 67.

for the satisfaction of their wants on the whole world, thus destroying the former natural exclusiveness of separate nations. It made natural science subservient to capital and took from the division of labour the last semblance of its natural character. It destroyed natural growth in general, as far as this is possible while labour exists, and resolved all natural relationships to money relationships. In place of naturally grown towns it created the modern, large industrial cities, which have sprung up over night.[1] Again the stress is on competition as that which divides classes internally—bourgeois and workers. But the growth of large-scale industry nonetheless is seen as providing the opportunity through which hitherto isolated individuals could unite against 'organized power' and overcome it after what Marx assumed would be a long struggle.

What Marx alludes to again and again are the various modes of domination which may exist on the basis of different forms of profit-making and be expressed in differing organizations of the division of labour. However, as against earlier forms of domination he stresses the artificial character of social relations under capitalism. He has in mind the impersonal character of exchange relationships. The 'natural' bonds of family, tribe, community are eradicated and replaced by the bond of all bonds, money; and large-scale industry implies the increasing differentiation (or fragmentation) of labour. It is the enormous power of the productive forces thus created over and against the producers who individually play such small parts in creating the totality. The distinctive character of industrial capitalism is brought out by Marx in the following passage which is certainly in line with his earlier remarks in *The Economic and Philosophical Manuscripts*, albeit with a stronger empirical anchorage (even if one chooses not to accept the inference he himself draws):

> Never, in any earlier period, have the productive forces taken on a form so indifferent to the intercourse of individuals *as* individuals, because their intercourse itself was formerly a restricted one. On the other hand, standing over against these productive forces, we have the majority of the individuals from whom these forces have been wrested away, and who, robbed

[1] K. Marx and F. Engels, *The German Ideology*, Lawrence & Wishart, 1965, p. 76.

thus of all real life-content, have become abstract individuals, but who are, however, only by this fact put into a position to enter into relation with one another *as individuals*.

The only connection which still links them with the productive forces and with their own existence—labour—has lost all semblance of self-activity and only sustains their life by stunting it.[1]

And the overcoming of this state of alienation is nothing less than the appropriation of the existing totality of productive forces. It is to be accomplished by the revolutionary unity of the proletariat to abolish private property, the division of labour, and hence their own subordination in one stroke. Previous substitutes for community, of which the state was the latest example, would be replaced by a real community when the revolutionary proletarians took over the conditions of their existence. It will no longer be an illusory community in which individuals' behaviour and life-chances are constrained by class, but one in which the free development and movement of individuals become a reality. The fragmenting effect of the division of labour which made individuals feel helpless and in the grip of chance would thus be overcome. The alien bond would bind no more.

Now this concern with the effects of private property and the division of labour is a continuing theme in all Marx's writings. In *The Communist Manifesto* (1848) there is the explicit commitment to abolish private (bourgeois) property and transform it into social property, that is property which has lost its class character. The freedom and independence which private property is said to generate among the population is energetically rejected: 'In bourgeois society capital is independent and has individuality, while the living person is dependent and has no individuality.'[2] What is at issue is how an association of men may emerge in which 'the free development of each is the condition for the free development of all'.[3] Given the built-in antagonism of labour

---

[1] K. Marx and F. Engels, *The German Ideology*, Lawrence & Wishart, 1965, pp. 82-3.
[2] *Communist Manifesto*. Foreign Languages Publishing House, Moscow, 1957, pp. 75-6.
[3] *ibid.*, p. 89.

and capital, this could never be achieved in a capitalist society whose impersonal and uncontrollable power is emphasized:

> Modern bourgeois society with its relations of production, of exchange of property, a society that has conjured up such gigantic means of production and of exchange, is like the sorcerer, who is no longer able to control the power of the netherworld whom he has called up by his spells.[1]

The wage labourers are again pictured as at the mercy of competitive forces they cannot control and are regarded as a commodity. The introduction of machine-based industry entails proletarian work which 'has lost all individual character and consequently all charm for the workmen. He becomes an appendage of the machine, and it is only the most simple, most monotonous, and most easily acquired knack, that is required of him.'[2]

Now it is worth recognizing that in each of these works of Marx there is a strong tendency to denote the position of the worker under capitalism as necessarily physically exhausting, a source of mental stagnation, permitting only subsistence wages, living in poverty and appalling social conditions. In *The Economic and Philosophical Manuscripts* for example we find this:

> Even the need for fresh air ceases for the worker. Man returns to a cave dwelling, which is now, however, contaminated with the pestilential breath of civilization, which he continues to occupy only *precariously*, it being for him an alien habitation which can be withdrawn from him any day—a place from which, if he does not pay, he can be thrown out any day. . . . Light, air, etc.—the simplest *animal* cleanliness—ceases to be a need for man. *Filth*, this stagnation and putrefaction of man—the *sewage of civilization* (speaking quite literally)—comes to be the *element of life* for him. Utter, *unnatural* neglect, putrefied nature, comes to be his *life-element*.[3]

These observations are extended empirically in *Capital* and, of course, characterize Engels' *Condition of the English Working*

---

[1] *Communist Manifesto*. Foreign Languages Publishing House, Moscow, 1957, pp. 56–7.
[2] *ibid.*, p. 59.   [3] op. cit., pp. 148–9.

Class in Britain in 1844.[1] In *Wage Labour and Capital* (1849) Marx also traces the interaction between the increasing differentiation of the division of labour and the continual improvement of machinery. This typically intensifies competition among workers for employment and eventually leads to wage contraction. This even happens, he argues, if there is a rapid growth of productive capital which could lead to a natural improvement in the lot of some workers.

But:

> To say that the most favourable condition for wage labour is the most rapid possible growth of productive capital is only to say that the more rapidly the working class increases and enlarges the power that is hostile to it, the wealth that does not belong to it and that rules over it, the more favourable will be the conditions under which it is allowed to labour anew at increasing bourgeois wealth, at enlarging the power of capital, content with forging for itself the golden chains by which the bourgeoisie drags it in its train.[2]

The situation in the longer term remains unsatisfactory, however, because of the overall replacement of men by machines, the lack of adequate alternative employment opportunities, and a growing intensity of competition for the jobs which remain. These arguments we find are worked through in even greater detail in *Capital* (1886). Here we find in his discussion of the General Law on Capitalist Accumulation the view expressed that the price of labour may keep on rising if the progress of capital accumulation is not interfered with.

> The rate of accumulation is the independent, not the dependent, variable; the rate of wages, the dependent, not the independent, variable. . . . The rise of wages . . . is confined within limits that not only leave intact the foundations of the capitalistic system, but also secure its reproduction on a progressive scale.[3]

---

[1] Allen & Unwin, 1892.
[2] K. Marx, 'Wage Labour and Capital' in *Selected Works of Marx and Engels*, Lawrence & Wishart, 1968, p. 88.
[3] K. Marx, *Capital*, Allen & Unwin, 1957, pp. 633-4.

And his discussion of surplus-value leads him to the following conclusion which echoes and amplifies his earlier comments in the *Communist Manifesto:*

> . . . within the capitalist system all methods for raising the social productiveness of labour are brought about at the cost of the individual labourer; all means for the development of production transform themselves into means of domination over, and exploitation of, the producers; they mutilate the labourer into a fragment of a man, degrade him to the level of an appendage of a machine, destroy every remnant of charm in his work, and turn it into hated toil, they estrange from him the intellectual potentialities of the labour process in the same proportion as science is incorporated in it as an independent power; they distort the conditions under which he works, subject him during the labour process to a despotism the more hateful for its meanness; they transform his lifetime into working time, and drag his wife and child beneath the wheels of the Juggernaut of the capital. But all methods for the production of surplus-value are at the same time methods of accumulation; and every extension of accumulation becomes again means for the development of those methods. It follows therefore that in the proportion as capital accumulates, the lot of the labourer be his payment high or low, must grow worse.[1]

What always appears to be fundamental in Marx's analysis is that it is only when capital itself is abolished that alienation will be abolished. From this perspective one can see that even granted the desirable improvement of wages and living standards, it would not suffice to 'solve' the alienation problem. Rather it would be the case of a limited good potentially getting in the way of a socialist community. It is only in this way that one can appreciate the acid comments which make up his *Critique of the Gotha Programme* (1885). The egalitarian wages programme, with its stress on 'fair distribution' cannot be considered independently of the mode of production, he argued. One still has a system of wage labour which 'is a system of slavery, and indeed of a slavery which becomes more severe in proportion as the social productive forces of labour develop whether the worker receives better or

[1] K. Marx, *Capital*, Allen & Unwin, 1957, pp. 660–1.

worse payment'.¹ To focus on redistribution as a primary factor is, in short, to take a retrogressive step. And, needless to say, even workers' cooperatives for the purposes of production could not in any final sense be justified if they were based on state aid: 'They are of value *only* insofar as they are the independent creations of the workers.'²

One detects a sense of unease in Marx's writings on the wages and standard of living issue (and, indeed, the potential growth in political participation of the working class in bourgeois society). It is embodied in what G. B. Shaw once called the capitalism of the proletariat. Daniel Bell has applied the phrase to account for much American trade union activity.³ In particular he points to the importance of the productivity wage increase. This wage strategy is credited in the first instance to General Motors who sought industrial peace by offering a five-year contract on this basis. Bell argues that trade union militancy has effectively diminished in the USA since the Second World War and that unions, while continuing to bargain, have accepted that there are built-in limits to what can be obtained. He is depicting what he alternatively described as market-unionism—in which a union responds to the 'realities' of the industrial environment in which it operates.

> In effect . . . the logic of market-unionism leads to a limited, uneasy partnership of union and company, or union and industry: uneasy because in many cases employers would still prefer to exercise sole power, although the more sophisticated employers know the value of such powerful allies as the unions in safeguarding their interests, uneasy too, because there is still the historic tendency of labour, acting as a social movement, to oppose the employers as a class.⁴

We may recall that in *Capital*, Marx noted another factor which could encourage capitalistic attitudes among workers, namely the piece-rate system of wages. It was a system which

---

[1] K. Marx, 'Critique of the Gotha Programme' in *Selected Works of Marx and Engels*, op. cit., p. 239.
[2] *ibid.*, p. 330.
[3] See D. Bell, 'The Capitalism of the Proletariat: a theory of American Trade Unionism' in *The End of Ideology*, Free Press, 1960.
[4] *ibid.*, p. 216.

could in principle reward skills, energy and staying power differentially, and to this extent could encourage individuality and a sense of independence and control over the work situation among workers. Manifestly, individuals' wages could be raised above the average. He argued that this could only intensify competition among labour itself and to this extent could not be beneficial to labour as a whole. But he could see its attraction to the individual worker and, indeed, appreciated that changes in the piece-rate or attempting to turn 'good' piece-rates into less good time rates on the part of the employer could lead in practice to constant battles between employer and employee. In other words, it was difficult always to convince workers that piece rates were against their best economic interests. A different aspect of the problem was piece work which involved the sub-contracting of labour. Under this system a worker accepts a price for a job from an employer and then recruits and supervises a group of labourers to accomplish it. This involved, in Marx's view, the exploitation of labour by labour in the service of capital. But, of course, in economic terms the contractor could do well out of it.

Now these possibilities could inhibit revolutionary activity on the part of the proletariat. They represent adjustments, so to speak, within the framework of alienation. But the matter does not rest there since there is some evidence to suggest that, despite Marx's analyses in various places—which depict the polarization of the two great classes as the prelude to the final conflict before the dictatorship of the proletariat is established to bring into being the socialist society—his own theory of surplus value implies the rise of a new middle class. Nicolaus, in the paper cited above,[1] draws attention to two linked works of Marx in German. These are the *Grundrisse* (1857–58) and the *Theorien* (1861–62)[2] (sometimes known as Volume IV of *Capital* although it pre-dates the first three volumes). Thus in *Theorien* he concedes that Malthus was right when he argued that unproductive consumers were necessary for a capitalist economy: 'Malthus' greatest hope—which he himself indicates as more or less utopian —is that the middle class will grow in size and that the working

---

[1] p. 146.
[2] K. Marx, Grundrisse der Kritik der politischer Ökonomie (Rohentwurf) (Dietz Verlag. Berlin, 1953), K. Marx, Theorien ueber den Mehrwert, Karl Kantsky, editor (Dietz, Stuttgart, 1919).

proletariat will make up a constantly decreasing proportion of the total population (even if it grows in absolute numbers). That in fact is the course of bourgeois society.'[1] And very explicitly in a comment on Ricardo, he states that the growth of productivity through mechanization will not only result in unemployment for some workers (the industrial reserve army of capital), but will turn some into a servant or middle class:

> What Ricardo forgets to emphasize is the constant increase of the middle classes, who stand in the middle between the workers on one side and the capitalists and landed proprietors on the other side, who are for the most part supported directly by revenue, who rest as a burden on the labouring foundations and who increase the social security and power of the upper ten thousand.[2]

Nicolaus argues, in the light of this, that one should not play the game of 'squeezing the concept of alienated labour hard enough to make all the categories of sociology, politics and economics come dripping out of it, as if this philosopher's touchstone were a lemon'.[3] If his analysis is correct, then the internal logic of Marx's argument does not lead inexorably to revolutionary conclusions. The service class will have its stake in the existing order. We may recollect here too Marx's comments on the nature of bureaucracy which, in a sense, although relatively undeveloped, mesh in with Nicolaus's exposition. Commenting on the structure of the East India Company Marx asks: 'Who then governs in fact under the name of Direction? A large staff of irresponsible secretaries, examiners, and clerks at India House. . . . The real court of directors and the real Home Government of India are the permanent and irresponsible *bureaucracy* "the creatures of the desk and the creatures of favour" residing in Leadenhall Street.'[4] While presented polemically, it is clear where Marx sees the allegiance of the servant class (or at least this segment of it) to lie. And it is interesting to note that in present-day

---

[1] Cited in Nicolaus, op. cit., p. 42.
[2] op. cit., p. 45. [3] op. cit., pp. 40–1.
[4] The Government of India in *New York Daily Tribune*, 20 July, 1853, cited in S. Avineri, *The Social and Political Thought of Karl Marx*, Cambridge University Press, 1968.

discussions of the service class, the conservative political orientation is generally stressed.[1]

What Nicolaus is in fact suggesting, perhaps, is that the solution to the problem of alienation is not as simple as some Marxists appear to assume. It can, however, be fairly added that this in itself need not lead to the abandonment of the concept. It might be argued, for example, that the transformation of human values into market values and the concern to abolish man's dominion over man is a continuing element in Marx's thought.

There remains another thorny problem, however, in considering Marx's use of the alienation concept. In one way or another (no doubt with differing emphases and inferences on the way through his writings) alienation is inseparably linked with the process of capitalist accumulation. Here is the source of domination and exploitation of man by man. But might it not be that capitalist accumulation is replaced by socialist accumulation in and through which domination might still be exercised? This of course was a possibility envisaged by Trotsky. And it has led to continuing critiques of Stalinism in particular and state socialism in general. Thus, for example, Djilas notably has argued that contemporary communism is the party of ownership and exploitation in Eastern Europe.[2] Coerced collectivization, forms of compulsory labour, and the absence of trade unions independent of the state may be found on the one hand, with the privileges of the party functionaries established on the other—housing and material possessions being reflected in one's social standing.

> Labour cannot be free in a society where all material goods are monopolized by one group . . . speaking in the abstract, the labour force taken as a whole, is a factor in total social production, the new ruling class with its material and political monopoly uses this factor almost to the same extent that it does other material goods and elements of production and treats it in the same way disregarding the human factor.[3]

---

[1] For a summary and further bibliography of this question, see R. Dahrendorf, 'The Service Class' in T. Burns (ed.), *Industrial Man*, Penguin, 1969. Dahrendorf assumes here, as elsewhere, that Marxist analysis implies the polarization of classes, and that the rise of the service class was unforeseen by Marx.
[2] See M. Djilas, *The New Class*, Allen & Unwin, 1967.
[3] *ibid.*, pp. 103-4.

But, as Robert Conquest reminds us in his introductory note to *The New Class*, Marx himself had recognized that what he termed 'Raw Communism' could become 'the general capitalist' and 'by systematically denying man's personality be no more than the expression of private property'.[1] What we are confronted with, of course, is the uncomfortable reality of 'the state' with all its connotations of command which simply cannot be reconciled to the Marxist vision of the positive community, the kingdom of freedom, a vision which in the end is essentially anarchist.[2]

(2) *Alienation, Occupational Role, and 'Bad Faith'*

Jean-Paul Sartre has offered another interpretation of man's experience of alienation—in his analysis of the human condition as being open to the possibility of bad faith.[3] It is his distinctive method to offer concrete examples. One is from the world of work:

> Let us consider this waiter in the café. His movement is quick and forward, a little too precise, a little too rapid. He comes towards the patrons with a step a little too quick. He bends forward a little too eagerly; his voice, his eyes, express an interest a little too solicitous for the order of the customer. Finally there he returns, trying to imitate in his walk the inflexible stiffness of some kind of automaton while carrying his tray with the recklessness of a tight-rope-walker by putting in a perpetually unstable, perpetually broken equilibrium which he perpetually re-establishes by a light movement of the arm and hand. All his behaviour seems to us a game. He applies himself to chaining his movements as if they were mechanisms,

[1] See M. Djilas, *The New Class*, Allen & Unwin, 1967, p. 7.
[2] On this see Robert Tucker, *The Marxian Revolutionary Idea*, Allen & Unwin, 1969.
[3] See especially Jean-Paul Sartre, *Being and Nothingness*, Methuen, 1969.

the one regulating the other; his gestures and even his voice seem to be mechanisms; he gives himself the quickness and pitiless rapidity of things. He is playing, he is amusing himself. But what is he playing? We need not watch long before we can explain it: he is playing at *being* a waiter in a café. There is nothing there to surprise us. The game is a kind of marking out and investigation. The child plays with his body in order to explore it, to take inventory of it; the waiter in the café plays with his condition in order to *realize* it. This obligation is not different from that which is imposed on all tradesmen. Their condition is wholly one of ceremony. The public demands of them that they realize it as a ceremony; there is the dance of the grocer, of the tailor, of the auctioneer, by which they endeavour to persuade their clientele that they are nothing but a grocer, an auctioneer, a tailor. A grocer who dreams is offensive to the buyer, because such a grocer is not wholly grocer. Society demands that he limits himself to his function as a grocer, just as the soldier at attention makes himself into a soldier-thing with a direct regard which does not see at all, which is no longer meant to see, since it is the rule and not the interest of the moment which determines the point he must fix his eyes on (the sight 'fixed at ten paces'). There are indeed many precautions to imprison a man in what he is, as if he lived in perpetual fear that he might escape from it, that he might break away and suddenly elude his condition.[1]

Here then is the notion of a man becoming imprisoned in his role. To realize finally the condition of his role would, so to speak, be a form of total captivity—it would be to define oneself by one's function, and to exercise no longer one's freedom to act differently. Further, one cannot be identical with a role model, such as the model of 'the good waiter'.

In vain do I fulfill the functions of a café waiter. I can be he only in the neutralized mode, as the actor is Hamlet, by mechanically making the typical gestures of my state and by aiming at myself as an imaginary café waiter through those gestures taken as an 'analogue'. What I attempt to realize is a being-in-itself of the café waiter, as if it were not just in my

[1] *Being and Nothingness*, p. 59.

power to confer their value and their urgency upon my duties and the rights of my position, as if it were not my free choice to get up each morning at five o'clock or to remain in bed, even though it meant getting fired. As if from the very fact that I sustain this role in existence I did not transcend it on every side, as if I did not constitute myself as one *beyond* my condition.[1]

From examples such as the kind given, Sartre seeks to persuade us that we may gain insight into the nature of human reality and consciousness.

If bad faith is possible, it is because it is an immediate permanent threat to every project of the human being; it is because consciousness conceals in its being a permanent risk of bad faith. The origin of this risk is the fact that the nature of consciousness simultaneously is to be what it is not and simultaneously to be what it is.[2]

Bad faith is, then, a way of describing self-estrangement and is seen as a continuing threat to personal integrity or freedom. One of the few sociologists to take up this aspect of Sartre's work has been P. Berger.[3] In particular he suggests that the Marxist notion of false consciousness (or reified consciousness) is closely related to the concept of bad faith. He is pointing to the fact that the social world is a human construction and that to regard it as something other than a human product (facts of nature, divine will, or whatever) is to give it an objective existence which will falsely make one define one's actions in terms of necessity or fate:

Reification implies that man is capable of forgetting his own authorship of the human world, and, further, that the dialectic between man the producer and his products is lost to consciousness. The reified world is, by definition, a dehumanized world. It is experienced by man as a strange facticity, an *opus alienum* over which he has no control rather than as the *opus proprium*

---

[1] *Being and Nothingness*, p. 60.
[2] *ibid.*, p. 70.
[3] See for example, *Invitation to Sociology*, Doubleday, 1963. *The Social Construction of Reality* (with T. Luckmann), Allen Lane, 1967. *The Social Reality of Religion*, Faber, 1969.

of his own productive activity. . . . The objectivity of the social world means that it confronts man as something outside of himself. The decisive question is whether he still retains the awareness that, however objectivated, the social world was made by men—and, therefore, can be remade by them.[1]

The institutional orders of society and particular roles may be reified. So far as the latter are concerned Berger and Luckmann observe:

> The sector of self-consciousness that has been objectified in the role is there also apprehended as an inevitable fate, for which the individual may disclaim responsibility. The paradigmatic formula for this kind of reification is the statement 'I have no choice in the matter. I have to act in this way because of my position'—as husband, father, general, archbishop, chairman of the board, gangster or hangman, as the case may be. This means that the reification of roles narrows the subjective distance that the individual may establish between himself and his role playing.[2]

The analysis of self-estrangement in Berger and Luckmann hinges on three possibilities:

(a) The non-awareness that one has reified the social world.
(b) The apprehension of reification as a modality of consciousness.
(c) Attempts to build upon that knowledge and de-reify the social world in general and one's own roles in particular.

For the sociologist there is the additional interest in understanding the social circumstances which favour de-reification.

At the level of role performance, however, the critical point would appear to be not whether the person is conforming to expectations, but whether he is playing his role knowingly or blindly. To play it knowingly is to make it a vehicle of our own decisions rather than an occasion for bad faith. We may at this point remind ourself of Goffman's development of the concept of role distance as a term to describe behaviour of individuals who

---
[1] *The Social Construction of Reality*, p. 106.   [2] *ibid.*, p. 108.

wish to express *themselves* by indicating detachment, sometimes even disdain, for the role they are performing. The individual is in effect signalling that he is something other than the role he is performing. He may be suggesting 'I know you think people who perform this role are a certain *type* of person but I am not to be typed in this way. I am me. And this behaviour (gestures, speech, stance, etc.) is a demonstration of my individuality.' Goffman indeed says that 'the individual is denying not the role but the notion of self that is implied in the role for all accepting performers'.[1] However, notwithstanding the attempt to express one's individuality, Goffman also points out that role distance itself may become part of a typical role and, to that extent, routinized. Hence it may no longer represent individual spontaneity or even individuality. Further, the role is taken as given. It is not a matter of choosing to opt out of it, to express one's freedom, which is implied in Sartre and Berger. To this extent, the operation of role distance may be seen as a secondary adjustment to the total situation (secondary in the sense that no fundamental attempt is being made to challenge the existing structure of domination). 'Sullenness, muttering, irony, joking and sarcasm may all allow one to show that something of oneself lies outside the constraints of the moment and outside the role within whose jurisdiction the moment occurs.'[2]

It could be argued that in this way one is attempting to get some elbow room, or breathing-space, within an overall constraining situation. The employment of techniques which facilitate role distance may assist in the maintenance of a given social order or institution rather than offer or promote alternative structures of activity. The possibility of role distance in behaviour indicates, perhaps, an area of tolerated deviance. It would seem to suggest that the attempt to exercise some degree of control over one's role performance carries accommodative rather than revolutionary implications. Goffman himself speaks of the functions of role distance and gives an extended illustration of this in an operating theatre context. Indeed, it then becomes clear that when a superordinate (in this case the chief surgeon) manifests role distance it can be employed as a technique controlling the behaviour of others, or of tension management in what might be a crisis situation. Thus:

[1] E. Goffman, 'Role Distance' in *Encounters*, Bobbs-Merrill, 1961, p. 108.
[2] *ibid.*, p. 114.

Given the conflict between selecting a subordinate and helping him maintain his poise, it is understandable that surgeons will employ joking as a means of negative sanction, so that it is difficult to determine whether the joke is a cover for the sanction, or the sanction a cover for the joke. In either case, some distance from usual surgical decorum is acquired.[1]

However, whether it is role distancing to facilitate control over others, or to express self not covered by the role, the legitimacy of the on-going system is not fundamentally challenged:

> ... The person who mutters, jokes or responds with sarcasm to what is happening in a situation is nevertheless going along with the prevailing definition of the situation—with whatever bad spirit. The curent system of activity tells us what situated roles will be in charge of the situation, but these roles at the same time provide a framework in which role distance can be expressed.[2]

Role distance then may be a form of personal freedom realized in spite of (and perhaps helping to maintain) an institutional order, rather than as a mode of consciousness aimed at subverting and transforming it.

The hiatus (or even discrepancy) between a consciousness of asserting oneself in one's role (vocational fulfilment) or over and against one's role within limits (role distancing) and that consciousness which might throw into sharp relief the precarious nature of social structures (which formerly looked like strong prisons) brings us back to an unsolved problem in Sartre's *Being and Nothingness*. There the non-alienated man fulfils himself in the agony of choice (presumably because it *is* agony, Sartre writes of man as being condemned to freedom, minimally the shadowy freedom of bad faith). It would be difficult to see expressions of job satisfaction as of much relevance here! But, perhaps of greater importance, the release from self-estrangement is essentially an individual matter of personal integrity. It does not seem to imply anything wider concerning class action, or political action, as a consequence of personal freedom or as a

---

[1] E. Goffman, 'Role Distance' in *Encounters*, Bobbs-Merrill, 1961, p. 122.
[2] *ibid.*, p. 113.

prerequisite for it. It would seem rather to offer some insight into 'the essential human condition' than to discuss whether some structures are more oppressive or repressive in their impact on the individual than others. To this extent the attempt to 'interiorise Marxism seems not to succeed. Berger, it is true, tries to handle the matter by arguing that bad faith is not a matter to be simply applied to the individual in isolation but that indeed 'society provides for the individual a gigantic mechanism by which he can hide himself from his own freedom'.[1] And in this sense society always has to be seen as an immense conspiracy in bad faith. This is the attempted sociological filling out of 'the essential human condition' argument. But Berger goes on to point out that matters do not rest there: 'Every social institution can be an alibi, an instrument of alienation from our freedom. But at least some institutions can become protective shields for the actions of free men.'[2] There, however, the matter does appear to rest. The liberation of consciousness (which we may regard as the opposite of self-estrangement) entails, says Berger, that we confront the human condition without comforting mystifications, and this we can only do if we recognize that the taken-for-granted world is not the only world there is. But in place of an account of how authentic existence might be realized in particular forms of social structure (as opposed to others) there is only the relapse into metaphor:

> Society provides us with warm, reasonably comfortable caves, in which we can huddle with our fellows, beating on the drums that drown out the howling hyenas of the surrounding darkness. 'Ecstasy' is the act of stepping outside the caves, alone, to face the night.[3]

So—since liberation must take place within society—the need would seem to be that we construct the form of society in which ecstasy is a real option; but what that society would be like is hidden under Berger's label of sociological humanism. We only know that this would beget an attitude of scepticism towards the conservative myths of the present and the utopian myths of the revolutionary.

For this reason we conclude that to speak of ecstasy as the

[1] *Invitation to Sociology*, op. cit., p. 145.   [2] *ibid.*, p. 145.   [3] *ibid.*, p. 150.

mode of confronting the essential human condition would seem to leave us with as vague a notion of the desirable form of our future industrial society as does Marx's vision of an ultimate communism.

(3) *Max Weber and the Issue of Alienation in Industrial Society*

The concept of alienation *per se* is not found very frequently in Weber's writings. There is, however, a very central interest in the concept of power and the structure of control through which power is exercised. And in his extended, though not often alluded to, discussion on the division of labour, the concept of appropriation is much in evidence. One may, he suggests, usefully classify the division of labour not simply by reference to its economic or technical significance, but also according to its social significance. This includes a classification based upon the way in which different resources which have potential economic value—labour, the material means of production, and the opportunities for profit from managerial functions—are appropriated.

Taking the question of labour appropriation, Weber notes four basic possibilities. It may be appropriated by the individual worker himself; by an owner who views labour as his property, as in the slave system of unfree labour; by a workers' organization; or it may not be appropriated at all—in the sense that labour is formally free and is the subject of a contractual relationship. It should be recognized, that this last is a formal category and in practice Weber acknowledged that 'free labour' could be substantively regulated in various ways, so that one might tend to move to an owner-type position on the one hand or a worker-control situation on the other. There could, therefore, be effective controls over recruitment, training opportunities, the mobility and dismissal of workers. He, in fact, goes on to make the very significant observation that the appropriation of jobs by workers, or of workers by owners, in certain respects leads to rather similar results:

> In the first place, the two tendencies are very generally formally related. This is true when appropriation of the workers by an owner coincides with appropriation of opportunities for jobs by a closed organization of workers, as has happened in the manor associations. In such cases it is natural that exploitation of labour services should, to a large extent, be stereotyped; hence, that work effort should be restricted and that the workers have little self-interest in the output. The result is generally a successful resistance of workers against any sort of technical innovation. But even where this does not occur, the fact that workers are appropriated by an owner means in practice that he is obliged to make use of this particular labour force. He is not in a position, like the modern factory manager, to select according to technical needs, but must utilize those he has without selection. This is particularly true of slave labour. Any attempt to exact performance from appropriated workers beyond that which has become traditionally established encounters traditional obstacles. These could only be overcome by the most ruthless methods, which are not without their danger from the point of view of the owner's self-interest, since they might undermine the traditionalistic bases of his authority. Hence almost universally the work effort of appropriated workers has shown a tendency to restriction. Even where, as was particularly true of Eastern Europe at the beginning of the modern age, this was broken by the power of the lords, the development of higher technical levels of production was impeded by the absence of the selective process and by the absence of any element of self-interest or risk taking on the part of the appropriated workers. When jobs have been formally appropriated by workers, the same result comes about even more rapidly.[1]

In practice, therefore, appropriation of one's labour by another is another way of talking about domination. But one can perhaps note that, insofar as the appropriation of one's own labour

[1] Max Weber, *Economy and Society* (edited by G. Roth and C. Wittich), Bedminster Press, 1968, pp. 129–30.

services implies control over one's own fate and work pattern, it presents a picture of a non-alienated state. The mediaeval craftsman is cited as an example by Weber and the autonomy he had in deploying his services is often treated as a bench mark in discussions of alienation.[1]

The appropriation of the material means of production is also open to four basic possibilities:

(1) Individual workers may so appropriate—Weber cites boatmen, carters, artisans, and peasants with smallholdings, as cases in point in certain historical circumstances.

(2) Workers' organizations may appropriate—these may vary from various craft groups to producers' co-operatives. The individual worker may, however, no longer be an owner.

(3) Owners may appropriate. The things they may appropriate might include land, water, sources of power, work premises, tools, machinery, apparatus and raw materials. Simply to give the list is to recognize that there may well be more than one owner.

(4) Regulating groups of third party character may appropriate. Such groups may regulate the economic activity of the production system but not use for itself the capital goods so created as a source of income. Instead it places the output at the disposal of its members. Towns or municipalities are cited as cases in point.

Weber stresses that while ownership of the material means of production may be of several different kinds, there are technical factors which separate the individual worker from ownership of the means of production. These include situations in which many workers are needed simultaneously or successively to complete the process or product; where sources of power can only be effectively exploited when workers are brought under unified control; where the techniques involved in the work process need continuous supervision of complementary activities; where the utilization of labour is only effectively achieved in large scale undertakings, which demand a co-ordinating and technically trained management; and, perhaps notably, a situation in which unified control over the means of production and raw materials makes it possible for labour to be subjected to stringent discipline, by which

[1] See for example, H. Arendt, *The Human Condition*, Doubleday Anchor, 1969 and T. Veblen, *The Instinct of Workmanship*, Norton, 1941.

approaches are made to controlling work speeds, product quality and effort standardization. These factors, of course, delineate the typical post-industrial revolution situation. Indeed, Weber explicitly comments:

> From an historical point of view, the expropriation of labour has arisen since the sixteenth century in an economy characterized by the progressive extensive and intensive expansion of the market system on the one hand, because of the sheer superiority and actual indispensability of a type of management oriented to the particular market situations, and on the other because of the structure of power relationships in the society.[1]

Although Marx touched upon the matter from time to time, Weber is far more explicit, one may suggest, in his emphasis upon the fact that *all* workers, not simply manual but clerical, technical and even managerial, may be separated from ownership of the means of production, and in this sense are expropriated. Very important is his stress on the role of bureaucratic administration as a form of control. This was not simply a matter of capitalist enterprise in its full-grown state but could be applied, more so, to socialist economies 'which would retain the expropriation of all workers and merely bring it to completion by the expropriation of the private owners'.[2] Weber also explicitly recognizes the potential divorce of ownership from control in modern business enterprises—both the general growth of shareholdings and the role of outside financiers and banking interests. This could lead to a paradox: the enterprises set up to achieve long-run profitability (the formal rationality of the organization) might not only meet resistance from inside the organization in terms of various substantive rationalities, but in fact come to be dominated by 'outside interests' whose maximizing strategies might not always be congruent with the particular organization they controlled. On this he maintains:

> The fact that such 'outside' interests can affect the mode of control over managerial positions, even and especially when the highest degree of *formal* rationality in their selection is attained,

[1] op. cit., p. 138. [2] op. cit., p. 139.

constitutes a further element of *substantive* irrationality specific to the modern economic order. These might be entirely private 'wealth' interests, or business interests which are oriented to ends having no connection whatsoever with the organization, or finally, pure gambling interest. By gaining control of shares, all of these can control the appointment of the managing personnel and, more important, the business policies imposed on this management. The influence exercised by speculative interests outside the producing organizations themselves of the market situation, especially that for capital goods, and thus on the orientation of the production of goods, is *one* of the sources of the phenomena known as the 'crises' of the modern market economy.[1]

The point to be underlined here is the lack of control of individual participants—sometimes including managers. In this sense the modern market economy is seen both as encouraging widespread appropriation of the means of production by particular interest groups over and against the individual worker, and as having an uncontrolled (or substantively irrational) element built in.

It is instructive at this point to recall that Marx by 1864 had noted the internal developments of capitalism which appeared to necessitate the divorce of ownership from control in industry. He writes in Volume III of *Capital* of the functioning capitalist as a 'mere manager, an administrator of other people's capital' and the owner of capital as a 'mere owner, a mere money capitalist'. The profit the owner receives is received in the form of interest 'as mere compensation for owning capital that now is entirely divorced from the function in the actual process of reproduction, just as this function in the person of the manager is divorced from the ownership of capital. Profit thus appears . . . as a mere appropriation of the surplus-labour of others, arising from the conversion of means of production to capital, that is, from their alienation *vis-à-vis* the actual producer, from their antithesis as another's property to every individual actually at work in production, from manager down to last-day labourer. . . . This is the abolition of the capitalist mode of production within the capitalist mode of production itself. . . . It is private production

[1] op. cit., p. 140.

without the control of private property.[1] As Avineri notes, the picture presented here is of the worker alienated from his labour and the capitalist alienated from his capital, and it is reasonable to point to this as an anticipation of Burnham's *Managerial Revolution* thesis.[2] There appears to be a difference between Marx and Weber as they observe the same phenomenon. For Weber it was the arbitrary way in which managers themselves could be dominated by outside interests. For Marx it was the effective control which propertyless managers could exercise that stood out. Subsequent empirical work would seem to suggest that both conditions may be located.[3]

What is of importance in this context is that Weber is concerned to delineate the forms of domination which may arise based upon the *control* of the means of production and of labour. However, his very concept of power implies resistance and from this perspective we are sensitized to consider counter-strategies of control or, as some might express it, secondary adjustments within a framework of alienation.

This leads us to a linked issue. In many parts of his work Weber discusses the phenomenon of bureaucratization in modern industrial societies.[4] In doing this he stresses: (1) that the notion of the worker separated from ownership of the means of production can properly be extended to other spheres of social life: the university teacher is separated from ownership of the means of knowledge, libraries and scientific equipment and research facilities; the soldier is separated from ownership of the means of warfare, his weapons, uniform and equipment; the official (whether in industrial, religious, social or state organizations) is separated from ownership of the means of administration. This fact seemed to Weber to be endemic to modern industrial life. Hence, when in his discussion of socialism he notes the communist vision of the abolition of man's dominion over man, he manifestly doubts that the prophecy could be fulfilled.

---

[1] Cited in Avineri, op. cit., p. 178.
[2] See J. Burnham, *The Managerial Revolution*, Penguin, 1962.
[3] See for example, P. Sargent Florence, *The Logic of British and American Industry*, Routledge & Kegan Paul, 1953. G. Kolko, *Wealth and Power in America*, Praeger, 1962. J. Child, *The Business Enterprise in Modern Industrial Society*, Collier Macmillan, 1969.
[4] See especially M. Weber, *Economy and Society:* edited by G. Roth and C. Wittich, Bedminster Press, 1968, Vol. 3, chapter XI.

(2) Within the sphere of industrial production Weber notes the mutually reinforcing character of bureaucratic organizations and machine technology in shaping factory discipline. Writing of Germany in 1918 he says:

> What characterizes our current situation is this, that private economy bound up with private bureaucratization and hence with the separation of the worker's tools of his trade, dominates the sphere of *industrial* production which has never before in history borne these two characteristics together on such a scale; and this process coincides with the establishment of mechanical production within the factory, thus with a local accumulation of labour on the same premises, enslavement to the machine and common working discipline within the machine shop or pit. It is the discipline which lends the contemporary mode of separation of worker from materials its particular stamp.[1]

(3) In certain respects, however, the question is whether societies designated capitalist or socialist were transcended by the phenomenon of bureaucratization. Modern industrial societies are predicated upon the existence of an administrative class. This administrative class, as we have seen, is not in an ownership position. The fact that individual members are separated from ownership of the means of administration and, in this respect, are in a parallel position to industrial workers, should not lead one to assume an identity of interests between these expropriated groups. Again, in the Germany of his own time, Weber noted the very great growth in clerical occupations 'who have to be *educated* in a quite definite way, and who therefore have a definite *class* character'. And of this class he writes:

> Nothing is further from this class than solidarity with the proletariat, from whom, indeed, they endeavour rather to differentiate themselves increasingly. In varying degrees, but noticeably, the same is true of many sub-classes among these clerks. They all strive at least for similar *class* qualities, be it for themselves, or for their children. A *uniform* trend to proletarianization is not in evidence today.[2]

[1] Max Weber, 'Socialism' in J. E. T. Eldridge (ed.), *Max Weber: The Interpretation of Social Reality*, Michael Joseph, 1971, pp. 200–201.
[2] 'Socialism', op. cit., p. 211.

In other words, any notion of all those who have been expropriated from the means of production (in whatever sphere) and administration rising up with some common consciousness of their alienation to form a new society is severely discounted by Weber. We may recall that Weber sometimes, in observing the phenomenon of bureaucratization in modern industrial society, writes of the dictatorship of the official (as against the socialist transitional conception of the dictatorship of the proletariat). This should not deflect us from remembering that the bureaucrat is himself a man under authority and discipline. Consequently, his response to bureaucratic rules may have something in common with the response of the worker to factory discipline.

In a socialist state, Weber generally assumed that, if anything, the importance of a specialized bureaucracy would be increased. But whether in a socialist or capitalist society the essential role of bureaucratic organizations was reflected in the fact that 'when those subject to bureaucratic control seek to escape the influence of the existing bureaucratic apparatus, this is normally possible only by creating an organization of their own which is equally subject to the process of bureaucratization.'[1]

There is a further dimension of Weber's work which provides us with insight into the alienation theme which we may here take up. In doing so we will confront, not for the first time, the Chinese Box character of the concept.

In *The Protestant Ethic and the Spirit of Capitalism*, it will be recalled, Weber writes of the religious foundations of worldly asceticism, referring in particular to the puritan idea of the calling in which rational, methodical work activity is seen as pleasing to God (who, after all, in labouring to produce the world in six days had set a pioneering example). Now, from one point of view, the fact that the work was performed to God's glory, gave meaning to the activity. As a form of rational economic asceticism, however, it exacted its own price. Weber stresses the way in which this asceticism turned with all its force against the spontaneous enjoyment of life and all that it had to offer:

Impulsive enjoyment of life, which leads away both from work

[1] Max Weber, *Theory of Social and Economic Organizations*, Free Press, 1964, p. 338.

in a calling and from religion, was as such the enemy of rational asceticism, whether in the form of seigneurial sports, or the enjoyment of the dance hall or the public house of the common man.[1]

And work in a calling was measured against the criterion of profitability. It promoted, even granted a spiritual foundation, a spirit of acquisition, which, as we have already seen, Marx had observed as forging its own bonds. Weber writes:

> The idea of a man's duty to his possessions, to which he subordinates himself as an obedient steward, or even as an acquisitive machine, bears with chilling weight on his life. The greater the possessions the heavier, if the ascetic attitude towards life stands the test, the feeling of responsibility for them, holding them undiminished for the glory of God and increasing them by restless effort.[2]

The immediate point to notice here is that this rational approach to life, which, after all, is the antithesis of capriciousness and lack of man's mastery over the natural world (hence his powerlessness), is essentially interpreted as a form of renunciation or, in Marx's terminology, self-estrangement, a loss of humanity. Secondly, insofar as the spirit of asceticism served to promote the modern economic order, it was the midwife of a form of domination:

> The puritan wanted to work in a calling: we are forced to do so ... the modern economic order is now bound to the technical and economic conditions of machine production which today determine the lives of all the individuals who are born into this mechanism, not only those directly concerned with economic acquisitions, with irresistible force. ... Since asceticism undertook to remodel the world and to work out its ideals in the world, material goods have gained an increasing and finally an inexorable power over the lives of the men as at no previous period in history.[3]

---

[1] Max Weber, *The Protestant Ethic and the Spirit of Capitalism*, Allen & Unwin, 1930, pp. 167–8.
[2] *ibid.*, p. 170.   [3] *ibid.*, p. 181.

In fact the sequence which Weber traces in the overall development of victorious capitalism contains the deep irony that activities pursued in a 'rational' fashion lead ultimately to a life of futility and meaninglessness. There is:

(*a*) the life-renunciating character of a spirituality whose goal is profit-making to God's glory.

(*b*) The secular replacement of the puritan work ethic by the utilitarian work ethic. Moral values are themselves judged against their usefulness in promoting economic success.

(*c*) More and more the emphasis placed upon techniques which may be correctly understood and employed for increasing wealth.

(*d*) The eventual divorce of this emphasis upon techniques from moral valuations, which leads to a barrenness of life of which the participants may or may not be conscious.

These themes are sharply raised at the end of Weber's *Protestant Ethic* essay:

> ... the idea of duty in one's calling prowls about in our lives like the ghost of dead religious beliefs. Where the fulfilment of the calling cannot directly be related to the highest spiritual and cultural values, or when, on the other hand, it need not be felt simply as economic compulsion, the individual generally abandons the attempt to justify it at all. The field of its highest development, in the United States, the pursuit of wealth, stripped of its religious and ethical meaning, tends to become associated with purely mundane passions, which often actually give it the character of sport.
>
> No one knows who will live in this cage in the future, or whether at the end of this tremendous development entirely new prophets will arise, or there will be a great re-birth of old ideas and ideals, or, if neither, mechanized petrification, embellished with a sort of convulsive self-importance. For of the last stage of this cultural development it might be truly said: 'specialists without spirit, sensualists without heart; this nullity imagines that it has attained a level of civilization never before achieved.'[1]

This pessimism (which, while not total, is certainly dominant) contrasts markedly, of course, with the Marxist perspective in which the hope of a resurrected humanity is more confidently held out.

---

[1] Max Weber, *The Protestant Ethic and the Spirit of Capitalism*, p. 182.

(4) Trade Unions and Bureaucratic Control

It might be argued, as writers such as Schumpeter and Galbraith have done,[1] that democracy in industrial societies, if it is to be realized at all, must be expressed in a situation of competition between bureaucratically organized interest groups. Leaving aside the important question as to what kind of competition actually takes place between interest groups in terms of effective power and accomplished actions, there is another significant question which relates to the internal organization of the bureaucracy. Putting it into the context of our discussion of alienation the question is: are we to regard the modern trade union as an organization over which the individual worker has no effective control, such that having been set up in his name it remains to dominate him, rather than express his real interest? To answer in the affirmative is to suggest that the masses are alienated from power which is exclusively exercised by an élitist governing group. And this of course was the contention expressed in Michels' 'iron law of oligarchy' and specifically applied by him to political parties and trade unions.[2] His study, first published in 1911, with its key theme—'who says organization says oligarchy'—has played a continual if controversial part in academic work and debate since then. Michel's research strategy is to take the apparently least favourable situation for his thesis and then to indicate how it still appears to hold:

> The study of the oligarchical manifestations in party life is most valuable and most decisive in its results when undertaken in relation to the revolutionary parties, for the reason that these parties ... in respect of origin and of progress, represent the negation of any such tendency and have actually come into existence out of opposition thereto. Thus the appearance of oligarchical phenomena in the very bosom of the revolutionary parties is a conclusive proof of the existence of immanent

---

[1] J. A. Schumpeter, *Capitalism, Socialism and Democracy*, Allen & Unwin, 1943. J. K. Galbraith, *American Capitalism*, Penguin, 1967.
[2] R. Michels, *Political Parties*, Collins Books, 1962.

oligarchical tendencies in every kind of human organization which strives for the attainment of definite ends.[3]

In discussing trade union organization Michels emphasises the increasing need for a bureaucratic organization staffed with technically competent men who become better educated, and separated in life-style, from those they represent. Leadership and authority tend to get increasingly centralized and it becomes possible for the leadership to initiate actions not approved of by the majority of their members. Indeed Michels held the oligarchical tendencies of trade unions to be more pronounced than in the sphere of political organization. It might be thought that union leaders coming from a working class background would identify with their membership and represent their true interests. But what actually happens, Michels maintains, is that the proletarian leader is assimilated into the existing order. 'What interest for them has now the dogma of social revolution? Their own social revolution has already been effected.'[2] Michels cites the example of English trade unionists entering into sliding-scale agreements in which wages were related to the selling price of the product, as a case of a technical organizational solution which played down class antagonisms. American trade union leaders are, if anything, more strongly indicted:

> ... not a few labour leaders are altogether in the hands of the capitalists. Being uneducated parvenus, they are extremely sensible to flattery. ... In many cases they are no more than paid servants of capital. ... Among the best organized unions there are some which enter into regular treaties with the capitalists in their respective branches of industry in order to exploit a consumer and to effect with the capitalist a friendly division of the spoils. In other cases, the leaders of a federation of trade unions, bribed by one group of employers, will organize strikes among the employees of another group. On the other hand, many strikes which are progressing favourably for the workers come abruptly to an end because the employers have made it worth the leaders' while to call the strike off.[3]

Despite Michels' comments on the conservative tendencies of bureaucratic organizations—administrations tending to become

[1] R. Michels, *Political Parties*, p. 50.   [2] *ibid.*, pp. 283-4.   [3] *ibid.*, p. 289.

ends in themselves rather than means to achieve the known ends of the membership—it would be incorrect to interpret his study as an anti-democratic tract. Rather he was concerned to destroy illusions and claims made in the name of democracy which were a charade. Moreover, one could put the question: how could the iron law of oligarchy be checked? Democracy as an ideology could certainly act as a check in its own right: the very act of labouring to produce a greater measure of democracy provides a force for criticism and control of oligarchical tendencies. The unmasking of illusions is part of such action. Perhaps most important is Michels' stress on the role of education as a factor for diminishing the gap between the leaders and the led, and increasing the capacity for the masses to exercise control in consequence:

> Taken in the mass, the poor are powerless and disarmed *vis-à-vis* their leaders. Their intellectual and cultural inferiority makes it impossible for them to see whither the leader is going, or to estimate in advance the significance of his actions. It is, consequently, the great task of social education to raise the intellectual level of the masses, so that they may be enabled within the limits of what is possible to counteract the oligarchical tendencies of the working class movement.[1]

The Webbs, writing about English trade unions before Michels, came to similar conclusions. Indeed they are cited by Michels as confirmatory evidence in his own study.[2] The trade unions presented for them 'an unrivalled field of observation as to the manner in which the working man copes with the problem of combining administrative efficiency with popular control'.[3] They argued that 'if a democracy means that everything which "concerns all should be decided by all" and that each citizen should enjoy an equal and identical share in the government, Trade Union history clearly indicates the inevitable result. Government by such contrivances as Rotation of Office, the Mass Meeting, the Referendum and Initiative, or the Delegate restricted by his Imperative Mandate, leads straight to either inefficiency and disintegration or to the uncontrolled dominance of a personal dictator or an expert bureaucracy'.[4] They pointed to the substitution of these features of

---

[1] R. Michels, *Political Parties*, p. 369.   [2] *ibid.*, p. 67.
[3] S. and B. Webb, *Industrial Democracy*, London, 1911, pp. v and vi.
[4] *ibid.*, p. 36.

primitive democracy by 'the typically modern form of democracy, the elected representative assembly, appointing and controlling an executive committee under whose direction the permanent official staff performs its work.'[1] They pointed at the same time to the 'extreme centralization of finance and policy which the Trade Union has found to be a condition of efficiency,[2] which was at the root of an apparent paradox: 'the constant tendency to a centralized and bureaucratic administration [which] is in the Trade Union world accepted, and even welcomed, by men who, in all other organizations to which they belong, are sturdy defenders of local autonomy'.[3]

Looking at the question of trade union administration in Britain much later, V. L. Allen[4] argues that the increasing size and complexity of functions together with the need for efficiency has led to a bureaucratic form of administration (in the Weberian sense) because in his view trade unions derive power from efficient organization. They have adopted the bureaucratic form of organization at the expense of 'self government'. He argues that this is scarcely surprising since 'the end of trade union activity is to protect and improve the general living standards of its members and not to provide workers with an exercise in self-government'.[5] In so far as they are 'delivering the goods' for their members however, Allen suggests that union leaders and executives have not deserted democratic principles, since they are, typically, genuinely representing their members' interests. For him it simply illustrates 'the tendency for democracy to have a preference for the authoritarian solution of important problems'.[6] For Allen, unrepresentative union leadership leads to membership dissatisfaction and decline. 'Membership fluctuates frequently and substantially in trade unions in response to the kind of service they are expected by their members to provide. Membership fluctuations, therefore, represent fairly clearly the changes in the conditions of trade unionism.'[7]

Lipset has suggested that the union career of John L. Lewis gives some support to Allen's view. In the late 1920s and early 1930s he followed conservative union policies in leading the United Mine

---

[1] S. and B. Webb, *Industrial Democracy*, p. 37.
[2] *ibid.*, p. 102.   [3] *ibid.*, p. 103.
[4] V. L. Allen, *Power in Trade Unions*, Longmans, 1954.
[5] *ibid.*, p. 15.   [6] *ibid.*, p. 25.   [7] *ibid.*, p. 206.

Workers of America. But confronted with a rapidly declining union membership and the growth of left-wing opposition, Lewis then proceeded to adopt a much more militant policy.[1] And C. Wright Mills has concurred in this judgment that labour leaders in the United States who fail to 'deliver the goods' are at risk.[2] At the same time he portrays successful union leaders as seeking to act within and as part of the national power élite. As union leaders they do not seek radically to transform the social structure but to achieve a more advantageous integration of their members into the existing framework:

> The drift their actions implement in terms of the largest projections, is a kind of 'precapitalist syndicalism from the top'. They seek, in the first instance, greater integration at the upper levels of the corporate economy rather than greater power at the lower level of the work hierarchy, for, in brief, it is the unexpressed desire of American labour leaders to join with owners and managers in running the corporate enterprise system and influencing decisively the political economy as a whole.[3]

Whether this is acting in their members' interests is perhaps not something that can be empirically easily ascertained. Wright Mills appears to draw the conclusion that effective control of such leaders by the membership is made more difficult because many of their actions are invisible to the membership. This in itself implies that membership allegiance is based on ignorance rather than knowledge and is perhaps not well described positively as satisfaction. There is rather a zone of indifference among the membership based on lack of interest or lack of knowledge which gives the union leader what in the strict sense is power without accountability. The argument appears to shift, therefore, to one in which, while the union leader's insecurity may be related to expressions of discontent from the membership, his security does not rest upon positive satisfaction with his work by the majority of members.

---

[1] S. M. Lipset, in his introductory essay to Michels' *Political Parties*, op. cit., p. 29.
[2] C. Wright Mills, 'The Labour Leaders and the Power Elite' in A. Kornhauser, R. Dubin and A. M. Ross (eds.), *Industrial Conflict*, McGraw-Hill, 1954.
[3] *ibid.*, pp. 151–2.

However, the matter may be seen to be further complicated by the fact that membership fluctuations can only in a very guarded sense be treated as an indicator of membership satisfaction or dissatisfaction. The fluctuations may be caused by quite other factors, such as the general level of employment and the particular fortunes of the industries in which the union operates. Furthermore, a union may have established negotiating rights in a firm or industry which make it very difficult for dissatisfied members to set up alternative arrangements. There are great problems in bringing into being breakaway unions, as the Pilkington glass workers have recently discovered, notwithstanding their expressed disenchantment with their official union, the N.U.G.M.W.[1]

The situation is made much more difficult by the general drift, certainly in the UK, towards a more rationalized trade union situation of fewer and larger groupings. There, in any case, agreements between trade unions, such as the Bridlington principles operating in the UK, are of the 'no poaching' variety and can effectively control reductions in union membership arising from discontent. Factors of this kind lead one to pose again the question what constitutes union democracy? They have led Martin to argue that union democracy hinges on the existence of factions within a union: 'union democracy exists when union executives are unable to prevent opposition factions from distributing propaganda and mobilizing electoral support. . . . The survival of faction limits Executive ability to disregard rank and file opinion by providing the *potential* means for its overthrow . . . faction is an indispensable sanction against leadership failure to respond to membership opinion.'[2] The empirical question then becomes: what are the determinants of effective opposition with a union? Martin lists a range of factors including, for example, decentralized collective bargaining arrangements, a decentralized trade union structure with sub-structural autonomy, extensive and constitutional power in the hands of lay members and an indirect electoral system, unions operating in the sphere of craft technology, well educated union members, and unions with a high level of membership participation. But he explicitly recognizes that much more

---

[1] For further examples see S. W. Lerner, *Breakaway Unions and the Small Trade Union*, Allen & Unwin, 1961.

[2] Roderick Martin, 'Union Democracy: an Explanatory Framework' in *Sociology*, II, No. 2, 1968.

empirical work needs to be done before an adequate theory of union democracy can be formulated.

Turner's recent discussion of trade union democracy emphasises that the position of the worker in Britain can vary considerably.[1] This may be schematically summarized in the following way:

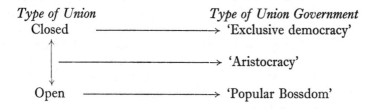

The closed unions in Turner's distinction are characterized by a restrictionist outlook which seeks to control the supply of labour to particular occupations and maintain an exclusive claim to employment within those occupations, whereas open unions are expansionist, relying on strength of numbers for their bargaining power. Turner discusses in detail the factors which both predispose a union to belong to one type rather than another, and which may lead it to move from an open to a closed position or the converse. 'Like most categories in the real world the borderline is blurred by mixed types and by transfer across it.'[2] The interesting point in this context is that the three types of union government can be located at different points on the closed/open continuum. The 'exclusive democracies' occur in closed union situations and are marked by: high membership participation in union affairs and management, and relatively few full-time officials—whose expertise and status is little different from that of the lay member (e.g. the London Typographical Society). The 'aristocracies' emerge where, essentially, one has, so to speak, first and second class citizens within a union—as, for example, when the Spinners recruited piecers, and the A.E.U. non-craft engineering operatives. The skilled men who maintained control over the supply of labour to their occupations constitute a lay aristocracy, filling to a disproportionate extent full-time as well as unpaid union positions. The 'popular bossdoms' exist in open union structures. Here we get a

---

[1] H. A. Turner, *Trade Union Growth, Structure and Policy*, Allen & Unwin, 1962.
[2] *ibid.*, p. 267.

low level of membership participation in union affairs (as Goldstein has illustrated with regard to the T. & G.W.U.),[1] a marked gap between the laity and the professionals in terms of expertise, and, commonly, a dominating General Secretary. Turner notes that in practice popular bossdoms may be characterized by different leadership styles: they may be radical or conservative, militant or conciliatory in their bargaining stances. This of course has different implications for the rank and file membership. The entrenched bureaucratic authority of the General Secretary and his District Officers can however be challenged despite apathy in many sectors; one of the most clear-cut examples of this is to be found in the chequered relations existing between the T. & G.W.U. and the dock workers.[2]

Notwithstanding Allen's earlier discussion which was an attempted partial refutation of Michels, he has more recently expressed views on the British trade union situation which essentially re-echo part of Michels' argument. Thus he points out that strikes are defined typically in the wider society as irresponsible, against the interests of the community, and so on.

> Union officials are particularly prone to the anti-strike environmental influences because they are frequently made out to be responsible for the behaviour of their members. Once they are committed to a strike call, union officials tend to become defensive, apologetic and concerned about taking avoiding action. When they are actually engaged in a strike they are frequently motivated by a desire to end it quickly irrespective of the merits of the issue.[3]

[1] J. Goldstein, *The Government of British Trade Unions*, Allen & Unwin, 1952.
[2] On this point see: V. L. Allen, *Trade Union Leadership*, Longmans, 1957 and J. Woodward et al., *The Dock Worker*, Liverpool University Press, 1954.
V. L. Allen, *Militant Trade Unionism*, Merlin Press, 1966, p. 27.

(5) *Alienation and Freedom: a Critique of the Blauner Thesis*

Robert Blauner's study *Alienation and Freedom* raises this question: under what conditions are the alienating tendencies of modern factory technology and work organization intensified, and under what conditions are they minimized and counteracted? The question is posed in the context of industrial life in modern America. What, here, is meant by alienation?

> Alienation is a general syndrome made up of a number of different objective conditions and subjective feelings and states which emerge from certain relationships between workers and the socio-technical settings of employment. Alienation exists when workers are unable to control their immediate work processes, to develop a sense of purpose and function which connects their jobs to the overall organization of production, to belong to integrated industrial communities and when they fail to become involved in the activity of work as a mode of personal self-expression. In modern industrial employment, control, purpose, social integration and self-involvement are all problematic.[1]

What we are offered from this perspective are four dimensions of alienation each of which may be contrasted with four non-alienative states:

1. Powerlessness ——————— Control
2. Meaninglessness ——————— Purpose
3. Isolation ——————— Social Integration
4. Self-estrangement ——————— Self-involvement

Blauner holds that what unites the four dimensions into the single concept of alienation is the fact that each of these expresses a principle of fragmentation which impedes the wholeness of man's experiences and activities. If one wishes to inquire into

[1] R. Blauner, *Alienation and Freedom*, University of Chicago Press, 1966, p. 15.

the alienation of the industrial worker, given adequate empirical data, one could comment on the *degree* of alienation of particular groups. This could be related to

(*a*) where the group was located on a particular dimension of alienation, and
(*b*) how the group appeared in profile when seen on all four dimensions.

The research strategy which Blauner adopts is, in essence, as follows: Given that the degree of alienation in its four dimensions is the dependent variable, he attempts to explain variations with reference to the type of technology as the independent variable—but also trying to take into account a number of other factors as intervening variables.

Empirical evidence from four industries based upon different technologies is considered:

1. Printing as an example of *craft technology*
2. Textiles as an example of *machine-tending technology*
3. Cars as an example of *assembly line technology*
4. Chemicals as an example of *process technology*.

Why does Blauner choose technology as his starting-point?

Variations in technology are of critical interest to students of the human meaning of work because technology, more than any other factor determines the nature of the job tasks performed by blue-collar employees and has an important effect upon a number of aspects of alienation. It is primarily the technological setting that influences the worker's powerlessness limiting or expanding the amount of freedom and control he exercises in his immediate work environment. Technological factors are paramount also in their impact on self-estrangement, since the machine system largely decides whether the worker can become directly engrossed in the activity of work or whether detachment or monotony more commonly result. Since technological considerations often determine the size of an industrial plant, they markedly influence the social atmosphere and degree of cohesion among the work force. Technology also structures the existence and form of work groups, in this way influencing cohesion. Even the nature of discipline and supervision to some extent depends on technological factors. And technology largely determines the occupational structure and skill distribution within an enterprise, the basic factors in advancement opportunities and normative integration.[1]

The three main intervening variables considered are: the division of labour, the social organization of the industry, and the economic structure of the industry. The first is depicted as being of particular relevance for the meaninglessness/purpose dimension and the second and third as influencing the isolation/social integration dimension. A summary of the form and nature of the inter-industry comparison is given in Chart II. The printer is presented as 'almost the prototype of the non-alienated worker in modern industry' and serves as a bench mark against which to express the social situation and experience of other industrial groups.

Blauner then uses the study as the basis for making a more ambitious statement about historical trends and future developments concerning worker alienation. He expresses the view that 'alienation has travelled a course that could be charted on a graph by means of an inverted U-curve'.[3] What he suggests is that in a craft technology situation the workers' freedom is at a maximum. This freedom is then sharply diminished by the advent of machine tending systems and is further intensified by mass assembly production systems.

> But with automated industry there is a countertrend, one that we can fortunately expect to become even more important in the future. The case of the continuous-process industries, particularly the chemical industry, shows that automation increases the worker's control over his work process and checks the further division of labour and growth of large factories. The result is meaningful work in a more cohesive, integrated industrial climate. The alienation curve begins to decline from its previous height as employees in automated industries gain a new dignity from responsibility and a sense of individual function—thus the inverted U.[4]

[1] R. Blauner, *Alienation and Freedom*, University of Chicago Press, 1966, p. 8.
[2] ibid., p. 57.   [3] ibid., p. 182.   [4] ibid., p. 182.

Blauner's American Inter-Industry Comparison

Type of Industry	Printing	Textiles
Type of Technology	Craft	Machine-tending
Degree of Alienation		
1. Powerlessness	Low	High
2. Meaninglessness	Low	High
3. Isolation	Low	Low
4. Self-Estrangement	Low	High
Nature of Explanation	Freedom and control largely due to nature of craft technology. This is strengthened by union power in the labour market and an upward growth in employment. Integration is fostered by a strong occupational community. Freedom from close supervision and an intrinsic interest in and satisfaction with the work.	Technology and division of labour produces powerlessness and the low status of the worker does not foster occupational identity. Meaninglessness is fostered by a highly specialized division of labour. But the worker achieves social integration in and through the local community where kinship and religion are dominant institutions.

Type of Industry	Automobiles	Chemicals
Type of Technology	Mass production assembly line	Process
Degree of Alienation		
1. Powerlessness	High	Low
2. Meaninglessness	High	Low
3. Isolation	High	Low
4. Self-Estrangement	High	Low
Nature of Explanation	Assembly line technology promotes meaninglessness over conditions of work and extreme division of labour takes meaning from the job. Lack of social integration is reflected in assembly line technology as inducing management labour conflict. Fluctuation in the product market promotes employment insecurity which mitigates against integration. The industry is usually located in anomic city environments. The job is regarded as a means to an end and not rewarding in itself as an activity.	Employment security and concept of work career for 'core labour force' in a buoyant long-term market for products. Process technology emphasizes responsibility and variety in work situation. Technology frees men from constant pressure. Social integration promoted through team production in decentralized plants. Identification with the enterprise is achieved through the work group. Work is self-actualizing because the dynamic nature of process technology is a challenge to the worker and prevents personal stagnation.

Remaining within Blauner's own frame of reference for the moment there are a number of caveats which may be entered, some of which Blauner himself recognized, though not always with sufficient emphasis. We may note the following:

1. The major source of inter-industry comparison is derived from a re-working of a Roper survey conducted in 1947 for *Fortune* magazine. In this respect it is certainly dated, although it is supplemented by case study material carried out by various investigators including Blauner himself in the case of a chemical company. The emphasis throughout is typically on expressed attitudes rather than on observed behaviour. The relationship between attitude and behaviour is of course one of the key problems confronting social science research.[1] Blauner acknowledges that he cannot testify to the representative nature of the case studies so far as the particular industries are concerned. And, of course, the four technologies chosen are illustrative rather than in any sense exhaustive so far as the employed population is concerned.

2. The relationship between industry and technology is not always clear cut. Blauner himself notes that in the American automobile industry only eighteen per cent of the manual work force were classified by the US Department of Labour in 1959 as actually working directly on the conveyor belt.[2] The 180 respondents from the industry in the Roper survey are not however differentiated, so it is difficult to get at the significance of assembly line work per se in assessing attitudes.

3. Obviously the predominant character of an industry's technology may change over a period of time. The case of printing is clearly a very important example of this. As an ideal type of the printing industry of a generation or so ago Blauner's example might serve, but it is doubtful whether it would apply to the dominant technology of the present-day printing industry. This point is however taken, that:

> ... technological innovations and economic developments threaten to eliminate not only the typesetter's control but the job itself. The newspaper industry has developed a process by which printed type can be set automatically by a columnist or

---

[1] For an incisive treatment of this issue see Percy S. Cohen, 'Social attitudes and Sociological Enquiry in *BJS*, XVII, No. 4, December 1966.
[2] See R. Blauner, op. cit., p. 91.

reporter as he writes out his copy on the typewriter. The craft unions may remain strong enough to resist this and other similar technological developments, but if they do not, printing may change rapidly from a craft to an automated industry.[1]

4. Great care should be taken in evaluating the inverted-U thesis of alienation even within Blauner's frame of reference. He is certainly not strictly inferring that automated forms of technology necessarily reduce alienation in the industrial work force. Thus data in, say, the oil industry, another example of continuous process technology, may not yield results identical to those of the chemical industry. His alienation trend thesis has to be set against his footnoted observation that 'automated technology will take many forms besides continuous-process production, and the diversified economic conditions of future automated industries will further complicate the situation'.[2] This last point ought perhaps to be emphasized. In the body of his argument Blauner maintained that, in capital-intensive industries, workers were under less pressure and had a greater sense of security than in labour-intensive situations. This rests mainly on the assumption that labour is treated as a fixed cost rather than a variable cost in production. The assumption may be queried. For example, the American oil industry, following the 1953-4 recession, instituted plans to cut labour costs—which involved the more efficient use of labour. In a situation of intensifying competition all costs, including those of labour, come under scrutiny. This, as Flanders has shown, was the impetus for productivity bargaining in the USA and in the UK.[3] At such times career patterns may be disrupted, expectations altered, and even redundancies contemplated. Part of the social integration story which Blauner tells in the chemical industry context, relates to the opportunities for upward mobility which this secure industry, with its formal job gradings and institutionalized mobility routes, offers. Yet he has to acknowledge, 'the economic downturn of the past few years' has sharply diminished opportunity for occupational advancement and in this respect has created 'a crisis of expectations'.[4]

When one looks at the evidence Blauner offers about the chemical industry—and his own case study of the Bay Chemical Company

[1] Blauner, op. cit., p. 57.   [2] op. cit., p. 182.
[3] See A. Flanders, *The Fawley Productivity Agreements*, Faber, 1964.
[4] Blauner, op. cit., p. 152.

consisted of only twenty-one interviews with blue-collar workers from production, maintenance, and distribution (seven each?)—the optimism of the thesis as baldly stated melts away. For example, he stresses the 'calm and crisis' pattern of work in chemicals which gives a pleasing unexpectedness to the job. It is seen as adding drama 'to what otherwise might be a pleasant but drab routine'.[1] Yet there are long stretches of time in which there is nothing to do, and complaints about monotony were common enough. And again the point is conceded that interest in the job and the intelligence needed for it should not be exaggerated. The activities of the operator typically consist of vigilance tasks. These, by their nature, are forms of clock-watching and must surely entail a heightened awareness of time. Yet this is embedded in Blauner's own definition of the self-estrangement dimension of alienation.

But there are criticisms of another order which must be raised.

1. In the first place, to write of the four dimensions of alienation is to suggest that each could be considered in terms of continuum. But more than this is seen to be involved. When, for example, Blauner analyses the notion of powerlessness, four distinct modes are differentiated: (*a*) separation from ownership of the means of production and the finished product (strictly these are two modes of powerlessness, not one), (*b*) inability to influence general managerial policies, (*c*) the lack of control over conditions of employment and (*d*) the lack of control over the immediate work process. The core of his analysis, however, is located in the last of these modes on the grounds that workers are most likely to value control over those matters which affect their immediate work tasks. He argues that American workers do not resent their lack of control in the first two spheres. He does not cite any evidence for this, but even if one assumes it to be so, the fact that these modes of powerlessness are taken as 'social givens' in the American context is implicitly to acknowledge the universality of alienation in that society. To restrict his attention to factors affecting control over the immediate work situation is simply to operate within the framework of alienation. Furthermore, although Blauner recognizes that lack of control over conditions of employment (the third mode of powerlessness) is not to be taken as a 'social given', he chooses not to develop variations in this sphere

[1] Blauner, op. cit., p. 156.

in any systematic way. This would appear to be a serious omission in a study which purports to comment on inter-industry differences in states of alienation. If employment regulation is taken as an indicator of control, then presumably the very effective power of the U.A.W. would have to be taken into account more seriously before the automobile workers were judged as the most highly alienated. None of this is to say that the things he does focus upon in the powerlessness dimension—ability to control the pace of work, to regulate the degree of pressure exerted over one by the work process, to maintain freedom of physical movement and freedom to control the quantity and quality of one's work and the way the work is actually carried out—are unimportant. The question of weighting the various elements thus subsumed is not really developed, but it is not unreasonable to suspect that the worker who was located at the freedom end of the continuum would express more satisfaction than one located at the other end. The real point, however, is that such results are not a sufficient basis for making general statements about the extent to which workers are alienated in industrial societies.

Secondly, if we turn to the isolation dimension we find that the other end of the continuum in normative integration of the individual is, so to speak, fragmented. This is because the sources of such integration might be variously the work group, the company, the industrial community, and the union, but one may readily recognize that empirically one form of integration (or kind of freedom) may reinforce another (e.g. the company and the community), or conversely they may be at the expense of one another (e.g. the work group versus the company). Hence, taking this dimension alone, many different kinds of behaviour might be described as non-alienated. This of course is the Durkheimian element in Blauner's analysis and it can potentially lead to confusion. For example, Kerr and Siegel's analysis of inter-industry differences in strike proneness, comments on the high strike proneness of geographically or socially isolated, cohesive homogenous groups of workers as a mark of their alienation from the wider society.[1] Yet their very cohesiveness is a mark of their normative integration as an occupational group.

[1] Clark Kerr and Abraham Siegel, 'The Interindustry Propensity to Strike— an International Comparison in Arthur Kornhauser, Robert Dubin and Arthur Ross (eds.), *Industrial Conflict*, McGraw-Hill, 1954.

I think another more crucial issue is also involved. In the light of the fact that each of the four dimensions is supposed to be an expression of the same general phenomenon, namely alienation, is it a matter of indifference as to which form of integration is discerned? The point is that some form of worker integration may actually mitigate against freedom on some modes of the powerlessness dimension. Indeed this came out explicitly in Blauner's own analysis of the textile worker. He describes the traditional mill village as one in which 'the accent is on personal relations with management, loyalties of kinship and neighbourhood, and a religious sanction on things as they are'. Kinship relations penetrate the work situation and Blauner goes on to say: 'Although such close family ties give the worker emotional security important in sustaining himself under difficult conditions of existence, they also inhibit striving for freedom and control. A worker who is self-assertive or joins a union may find that not only he but other members of his family will lose their jobs ... partly because of tradition and partly because of the submissive personalities of the workers. Mill managements still retain considerable paternalistic control.'[1] The logical status of the interrelation between the dimensions of alienation is therefore problematic. One might suggest that high integration be taken as a relevant consideration only insofar as it contributed to some component of control, and not when it served to entrench a form of domination.

The immediate significance of this conceptual point could well be that the inverted-U curve might need further revision: textile workers might be more alienated than automobile workers. Since their social integration in the community is at the expense of control in any of its modes they exhibit less awareness of their alienation than the car workers who attempt in some measure to alleviate their condition by militant action or effective collective bargaining. In other words the attempt to smuggle in a cure for anomie (social integration) as a cure for alienation (and without any of the radical connotations concerning social equality which we have attributed to Durkheim's treatment) is a spurious solution.

2. The second major query is to ask whether it was strategically sound to select the type of technology as the major independent variable in explaining the degree of alienation of industrial workers. If car workers, for example, are defined as having

[1] Blauner, op. cit., pp. 75–7, *passim*.

'instrumental attitudes'—staying in the industry for the high wages and putting up with the monotony, boredom and pressures of the assembly-line, one might properly wonder what there was about the wider society and their structural location in it which encouraged them to act as economic men in spite of the very real deprivations of the work situation.[1]

3. It is instructive to recall that Wright Mills' treatment of the alienation theme with regard to American society sounded a much more pessimistic note than Blauner. One important reason for this is to be found in the way Mills handles the distinction between the objective (structural factors) and subjective (feeling states) aspects. Blauner, in effect, concentrates on subjective feelings of powerlessness, meaninglessness, normlessness and isolation in assessing the degrees of alienation and then seeks to account for these by reference back to structural factors, and notably, as we have seen, the nature of technology. This means in practice that expressions of satisfaction are treated as positive signs of non alienation. By the same token, this avoids (some might say solves) the issue of false consciousness in Marxist literature which is employed to explain what otherwise cannot be explained in a somewhat arbitrary way—the Marxist equivalent of 'the God of the gaps' appealed to by some Christian apologists. However, if the sociologist wants to argue that social structures affect people's lives in ways which the people themselves only partially comprehend, then the actor's definition of the situation, whilst of great importance in sociological analysis, does not exhaust the sociologist's task. It is this guiding thought which shapes Mill's assessment of the notion of satisfaction and the commonly allied concept of morale as revealed in the following formulation:

> The theoretical problem of industrial sociology, as it comes to an intellectual and political climax in the conception of morale, is a problem of exploring the several types of alienation and morale which we come upon as we consider systematically the structure of power and its meanings for the individual lives of workmen. It requires us to examine the extent to which psychological shifts have accompanied structural shifts; and in each

[1] See for example John H. Goldthorpe, 'Attitudes and Behaviour of Car Assembly Workers: a deviant case and a theoretical critique' in *BJS*, XVII, September 1966.

G*

case why. In such directions lies the promise of a social science of modern man's working life.[1]

In his major substantive work in this area, *White Collar*,[2] Mills takes as a bench mark of non-alienation an ideal typical formulation of craftsmanship which includes the following elements:

(1) There is gratification in work, in the product that is made and the processes of its creation.

(2) There is satisfaction in the final product and the anticipation of its completion makes the details of his daily work meaningful to the craftsman.

(3) The craftsman is free to control his work activities such that there is a unity between the planning and performance of his work.

(4) In his work the craftsman develops his capabilities and skills: 'in this simple sense he lives in and through his work, which confesses and reveals him to the world'.[3]

(5) There is no separation of work and play or work and culture. The craftsman in his activity fuses work and play.

(6) There is no artificial separation between work and leisure: 'The craftsman's work is the mainspring of the only life he knows: he does not flee from work into a separate sphere of leisure: he brings to his non-working hours the values and qualities developed and employed in his working time.[4]

Mills goes on to indicate the structural factors which have led to the erosion of this craftsmanship model. Basically he elaborates factors already noted by Marx and Weber: urbanization and the growth of large-scale industry, the legal framework of modern capitalism, the detailed division of labour and the character of bureaucratic enterprises. Not only is the model of craftsmanship objectively unavailable for structural reasons, one cannot assume that it figures in the consciousness of workers in terms of a state of frustration, deprivation or rebellion as the case may be, because the model is not objectively attainable:

We cannot compare the idealized portrait of the craftsman with that of the auto worker and on that basis impute any psycho-

[1] C. W. Mills, *The Sociological Imagination*, O.U.P., 1959, p. 95.
[2] C. W. Mills, *White Collar*, O.U.P., 1951. [3] *ibid.*, p. 222. [4] *ibid.*, p. 223.

logical state to the auto worker. We cannot fruitfully compare the psychological condition of the old merchant's assistant with the modern saleslady, or the old-fashioned bookkeeper with the IBM machine attendant. For the historical destruction of craftsmanship and of the old office does not enter the consciousness of the modern wage-worker or white collar employee; much less is their absence felt by him as a crisis as it might have been if in the course of the last generation, his father or mother had been in the craft condition—but statistically speaking, they have not been.... Only the psychological imagination of the historian makes it possible to write of such comparisons as if they were of psychological import. The craft life would be immediately available as a fact of consciousness only if in the lifetime of the modern employees they had experienced a shift from one condition to another, which they have not; or if they had grasped it as an ideal meaning of work, which they have not.[1]

In *White Collar*, Mills does not discount the fact that workers may express satisfaction with their jobs but insists that the sources of the satisfaction are of a different order from the craftsmanship ideal. They are grounded in considerations of income, security, power and status. But this has nothing intrinsically to do, in Mills' view, with the full development of activity: indeed the satisfaction thus expressed may be nothing other than 'the morale of the cheerful robots'. The basis for this contention derives from his conviction that satisfaction may be engineered by a managerial élite. The human relations ideology on this reckoning exists to promote control over employees by manipulation rather than directly coercive techniques. But by unmasking the élitist character of the control structure and the manipulative manner in which satisfaction can be engendered Mills believes that here are realistic grounds for talking about false consciousness. This would appear to be the significance of the following comment:

> Current managerial attempts to create job enthusiasm ... are attempts to conquer work alienation within the bounds of work alienation. In the meantime, whatever satisfaction alienated men gain from work occurs within the framework of alienation.[2]

[1] C. W. Mills, *White Collar*, O.U.P., 1959, pp. 227-8   [2] *ibid.*, p. 235.

Part Four
Conclusion

What kind of integration?

If sociologists are at root fascinated with the question: how is society possible?—then in one shape or form they go on to explore the relationship of the individual to society and the interconnection between the various 'parts' of society. These questions are raised in an acute form when we look at industrial societies—both because of their institutional complexity and the rapid social, technological and scientific changes which characterize them. In a way, much of the concern of this book has centred upon the problem of integration in industrial societies, but in saying that there are certain points which should be made very clear.

Approaches to integration—the social integration of the individual into society and the integration of the various institutional orders in society—can and have taken many different forms in industrial societies. The major experiments to date include liberal laissez-faire 'market' societies, fascist and communist 'planned' societies, and 'welfare state' societies. There are in principle questions which can be raised about the efficacy of certain modes of integration in particular societies and other questions of a political nature when one discovers that a society is, or is not, well integrated.

At this point it may be as well to draw attention to the implications of Dahrendorf's comments on para-theory so far as our own discussion is concerned.[1] He reminds us that many of the great debates in European sociology lead to argument rather than empirical testing and that surrounding technical sociological investigations are matters of moral, political and philosophical import. It is not only, however, that they surround the sociologist and his work, they impinge on the concepts he employs—because his concepts are affected by his theory of society. Dahrendorf applies this judgement to his own writing:

[1] See R. Dahrendorf, *Essays in the Theory of Society*, Routledge, 1968.

Power, resistance, conflict, historical change, openness, freedom, uncertainty—with varying emphasis these notions pervade the present volume. Inevitably the sociological perspective made up by such notions is a political and philosophical perspective as well. It is the perspective of a modern liberalism: averse to utopia in sociological theory as in political practice (and in that sense neither 'conservative' nor 'radical') always looking for ways to guarantee individual liberty in a world of constraints, confident in the ability of the right social and political institutions to provide the possibility of free human development, and skeptical of all theories and approaches in social science that ignore or neglect the question of what they have to contribute to bringing about a society in which men may be free.[1]

As it happens, we have a good illustration of what is involved here in another work of Dahrendorf's, *Society and Democracy in Germany*. There we find him arguing that it is important to distinguish between social integration and social harmony:

> The confusion is dangerous because it suggests that creating harmony is the first task of politics; but this can never be accomplished except by repression. The confusion is erroneous because there may be well-integrated communities in which lively conflicts take place. . . .[3]

Yet, in Dahrendorf's view this confusion has existed in German political and industrial life. There is a preoccupation with avoiding conflict and with establishing coalitions in politics and co-partnership in industry. He writes of the institutionalized fear of social conflict which is notably reflected in the industrial sphere. Hence, for him, the dislike of industrial disputes in Germany is not altogether desirable:

> It is a symptom of an unfortunately rigid and unimaginative attitude towards the adversities of industrial life that is often coupled with utopian hopes and is for that very reason unprepared for real problems.[4]

---

[1] R. Dahrendorf, *Essays in the Theory of Society*, pp. vii–ix.
[2] R. Dahrendorf, *Society and Democracy in Germany*, Weidenfeld-Nicholson, 1968.
[3] *ibid.*, p. 195.   [4] *ibid.*, p. 178.

The unpreparedness is revealed in the relative lack of collective bargaining and negotiating procedures of an autonomous kind between employers and employees. His thesis is that no matter how noble the motives of the utopians who seek harmony, the result is a tendency towards illiberalism. Conflict is repressed, but notwithstanding the low strike statistics, it is not eliminated—rather it is reflected in individual not collective responses:

> Thus we find instead of work disputes, individual actions whose connection with social conflict is barely regonisable at first sight. Sinking work morale, growing fluctuation, indeed even sickness and accident rates may be indicators of such redirections of industrial conflict.[1]

So, for Dahrendorf, integration is to be understood within a pluralist framework in which adequate institutions for conflict regulation are established as between the State on the one hand and the enterprise on the other, so far as industrial life is concerned.

It is within the context of Dahrendorf's modern liberalism therefore that his usage and evaluation of the concept integration is to be understood. However, one cannot place restrictions on who uses words, and to what end and so one finds some radical sociologists placing a negative value on certain forms of integration. This occurs when they interpret integration as an outcrop of élite domination in industrial societies. From that perspective integration is not the polar opposite to alienation but an expression of an alienated condition. We have already encountered this view in Wright Mills. He was, as we have seen, critical of the human relations ideology which sought to create job satisfaction because he saw such attempts as taking place within the existing power structure and not truly enhancing human freedom. Rather it was the programming of cheerful robots. But this negative evaluation can only be fully appreciated when one recalls that his view of American industrial society was one in which the great bulk of the population were politically powerless against an anonymous, impersonal and highly centralized system of control:

> As political power has been centralized, the issues professionalized and compromised by the two-party state, a sort of imper-

[1] R. Dahrendorf, *Society and Democracy in Germany*, p. 178.

sonal manifestation has replaced authority. For authority there is a need of justifications in order to secure loyalties: for manipulation there is exercise of power without explicit justifications, for decisions are hidden. Manipulation . . . arises when there is centralization of power that is not publicly justified and those who have it don't believe they could justify it. Manipulation feeds upon and is fed by mass indifference. For in the narrowed range of assertion and counter-assertion no target of demand, no symbols or principles are argued over and debated in public. If areas of assertion and counter-assertion are narrow in the mass media, it is in some part because politics is monopolized by the two major parties and the economic-political arena of struggle, monopolized by the labour-union—corporation battle.[1]

A similar preoccupation with integration as a product of manipulative, anonymous élite power in industrial societies is found in the influential writings of Herbert Marcuse. Indeed in *One Dimensional Man* his intellectual debt to Wright Mills is explicitly acknowledged.[2] In a very different vein from Kerr and his associates, whose position we discussed in Part I, Marcuse attempts a characterization of all industrial societies, probing behind such labels as liberal, welfare, socialist or capitalist. He focuses upon the significance of technological development: with its attendant features of increased productivity, efficiency and the general rise in the standard of living. Yet the organization of production, consumption and distribution in advanced industrial societies is one of unreason and unfreedom. It leads him to label all such societies as totalitarian:

> By virtue of the way it has organized its technological base, contemporary industrial society tends to be totalitarian. For 'totalitarian' is not only a terroristic political co-ordination of society, but also a non-terroristic economic-technical co-ordination which operates through the manipulation of needs by vested interests. It thus precludes the emergence of an effective opposition against the whole. Not only a specific form

[1] *White Collar*, op. cit., p. 349.
[2] The following references are relevant to the issues now to be noted: H. Marcuse, *Eros and Civilisation*, Sphere Books, 1969. *One Dimensional Man*, Routledge, 1964. 'Liberation from the Affluent Society', *in* D. Cooper (ed.), *The Dialectics of Liberation*, Penguin, 1968.

of government or party rule makes for totalitarianism, but also a specific system of production and distribution which may well be compatible with a 'pluralism' of parties, newspapers, countervailing powers, etc. Today political power asserts itself through its power over the machine process and over the technical organization of the apparatus.[1]

None of this means that important qualitative distinctions between different forms of totalitarianism might not be made and in this case between the kind of society which is maintained by overt and forceful domination and that which is maintained in an administered society. Integration into the second kind is still a 'fateful' matter for Marcuse because key decisions affecting personal and national security are made and the individual has no control over them. He lives therefore in servitude. No matter what his material comforts may be in it, the administered society is a repressive society and the individual is therefore in a state of alienation. On this reading the Welfare State is interpreted as 'a system of subdued pluralism, in which the competing institutions concur in solidifying the power of the whole over the individual. . . . The reality of pluralism becomes ideological, deceptive. It seems to extend rather than reduce manipulation and co-ordination, to promote rather than counteract the fateful integration'.[2] From this perspective pluralism is more beguiling than authoritarianism in the sense that it is more plausibly a situation of individual freedom, yet in the name of freedom, democracy is a system of domination.

One of the major components of Marcuse's para-theory is that one may see individuals in advanced industrial civilization integrated into a social reality (which provides them with satisfaction) yet the reality represents a state of alienation. This is because despite all the rational characteristics of industrial societies as represented in the systems of production, exchange and distribution, there is a higher irrationality. This irrationality resides in the fact that all the promises of technological development are as nothing because technology is developed for destructive and warlike purposes. In this sense technological rationality sins against Reason.

In his writings Marcuse fluctuates somewhat between optimism

[1] *One Dimensional Man*, p. 3. [2] *ibid.*, pp. 50–1.

and pessimism, but it is important to recognize that 'liberation' from a state of alienation does not involve a rejection of technology and industrial life as a material basis of an alternative society. Indeed, automation carried to its logical possibilities he sees as providing the basis for the Marxist vision of the 'abolition of labour':

> Complete automation in the realm of necessity would open the discussion of free time as the one in which man's private *and* societal existence would constitute itself. This would be the historical transcendence toward a new civilization.[1]

The clue to what Marcuse understands by a new civilization resides in what he terms 'the pacification of existence'. It is used to designate

> the historical alternative of a world which—through an international conflict which transforms and suspends the contradictions within the established societies—advances on the brink of a global war. 'Pacification of existence' means the development of man's struggle with man and with nature under the conditions where the competing needs, desires and aspirations are no longer organized by vested interests in domination and scarcity—an organization which perpetuates the destructive forces of this struggle.[2]

In Marcuse's view, technological development creates the material basis for the pacification of existence. However, the vested interests of capital, labour and the political parties seek to contain rather than realize this possibility. It is this which leads Marcuse to write of a fateful integration. It is not, we may notice, a total integration even in its own terms: Marcuse refers to the inhuman existence of the poor, the unemployed and unemployable, the persecuted coloured races, the inmates of prisons and mental institutions, and appears to treat this as a necessary cost of affluence in which private appropriation and distribution of profit as an economic regulator are maintained. Certainly they prove a threat to the system:

[1] *One Dimensional Man*, p. 37.   [2] *ibid.*, p. 16.

They exist outside the democratic process: their life is the most immediate and real need for ending intolerable conditions and institutions. Thus their opposition is revolutionary even if their consciousness is not. Their opposition hits the system from without and is therefore not deflected by the system: it is an elementary force which violates the rules of the game, and in doing so, reveals it as a rigged game. When they get together and go out into the streets, without arms, without protection, in order to ask for the most primitive civil rights, they know that they face dogs, stones, and bombs, jail, concentration camps, even death. Their force is behind every political demonstration for the victims of law and order. The fact that they start refusing to play the game may be the fact which marks the beginning of the end of a period.[1]

The slenderness of the chance that this might be the basis on which a pacified social life might be inaugurated is indicated.

MacIntyre, in a sharp critique, has suggested that Marcuse shares with writers such as Bell and Lipset the end of ideology doctrine to the effect that changes in the structure of the labour force and consumption patterns, together with the growth of welfare state institutions have eroded traditional conflicts between labour and capital.[2] Very different however is the mood. It is not to be seen as an academic Puritan reacting to the fact that more people, including the working classes, can enjoy a better standard of living: rather it is the conviction that this takes place within a context in which nature is despoiled, used as an 'instrument of destructive productivity' and science and technology applied to military ends. We do not find in those other writers any pronounced concern with social reconstruction—certainly not with the pacification of existence. The possibility of its realization (and Marcuse grows a little in confidence concerning its realization in his later writings) explains his fury with the prosperous warfare and welfare state,

Comfort, business and job security in a society which prepares

[1] *One Dimensional Man*, pp. 256–7.
[2] See A. MacIntyre, *Marcuse*, Fontana, 1970. The studies referred to are D. Bell, *The End of Ideology*, op. cit., and S. M. Lipset *Political Man*, Mercury, 1963.

itself for and against nuclear destruction may serve as a universal example of enslaving contentment.[1]

Written into Marcuse's analysis as we have seen is a model of élite domination. An important part of his vision of an alternative society is that the pacification of existence implies a diffusion of power:

> Peace and power, freedom and power, Eros and power may well be contraries ... the reconstruction of the material base of society with a view to pacification may involve a qualitative as well as a quantitative *reduction* of power in order to create the space and time for the development of productivity under self-determined incentives.[2]

Here, of course, we see a very different basis for social integration. It should serve to remind us, by the same token, that integration does not have to be interpreted as negotiating adjustments to the status quo. It is, for example, different from the situation in the Soviet Union, recently discussed by Gouldner, in which he suggests that the sociological task is shaped by technological imperatives. There is

> ... the tacit assumption that one takes as *given* the jobs that have to be done, the basic social roles for which people have to be prepared, and the basic institutions in which they have to operate. For this reason there has been a growing interest in the sociology of industry.... But the Soviet sociology of industry is concerned with only a special case of the larger task, as one man puts it, of 'adjusting expectations to reality'. Nothing could make it clearer than this statement does that Soviet sociology, like Western functionalism, takes certain parts of its social world as given and views its mission as making them work together more smoothly. The problem is one of getting men to fit into and accept social institutions that are treated largely as givens. To conceive of integration as a problem of 'proportions' is to conceive of it in terms of a model which views the system as having 'requisites' and 'parts' that remain

[1] A. MacIntyre, *Marcuse*, p. 243. [2] *One Dimensional Man*, pp. 235-6.

essentially stable, though their links with one another may be strengthened or modified.[1]

We have thus tried to illustrate some of the ways in which sociologists have conceptualized the notion of integration. This has taken us away from the relatively straightforward question of what the empirical data tell us about this or that aspect of industrial life and into the realm of 'para-theory'. None of this is intended to suggest that one should ignore or become careless about the empirical questions. We may notice, for example, that MacIntyre attacks Marcuse's *One Dimensional Man* partly on empirical grounds. To speak of advanced industrial societies as highly integrated is very questionable. Technological advance and investment affects the various sections of the social order differentially:

> The result is not the highly integrated and well co-ordinated system portrayed by Marcuse, but rather a situation in which there is less and less co-ordination between different sectors.... The feeling of impotence that many have is not misplaced. They are impotent. But they are not impotent because they are dominated by a well-organized system of social control. It is lack of control which is at the heart of the social order and governments reflect this impotence as clearly as anyone else.[2]

As it happens, in the case we have cited, MacIntyre attempts to refute Marcuse empirically but he only resorts to counter-assertion in practice. The scientific character of sociology, however, can only be retained if such empirical matters are properly inquired into. But I have drawn attention to MacIntyre's comment for another reason. Broadly, our discussion of alienation has focused on the question of domination and control, and, as we have seen, a ruling class or élite theory has frequently been written into the analysis. The discussion of anomie, on the other hand, has centred on the question of fragmentation and regulative breakdown in the social order. Both of these concepts have a sensitizing function in helping us to understand the connections and the discontinuities which may be discerned between industry and the rest of social

---

[1] A. Gouldner, *The Coming Crisis of Western Sociology*, Heinemann, 1971, p. 466.
[2] *Marcuse*, op. cit., pp. 70-1.

life. As concepts they have reference to different kinds of empirical processes. In their various ways, classical and contemporary sociologists have sought to bring a new awareness of what is going on in industrial societies: both by the construction of such sensitizing concepts and by the conduct of careful empirical studies. Critics of sociology no doubt have many weapons in their armoury: they can castigate its practitioners as idle and irrelevant speculators or as mere fact-grubbers according to taste. One thing, however, may be said: many sociologists, whilst properly concerned with scientific method and the suspension of prejudice which that implies, have in terms of their own philosophical positions been intensely preoccupied with the conditions in which individual and social freedom may be established in industrial societies. It is a form of humanism which, in Tawney's words, embodies

> . . . the belief that the machinery of existence—property and material wealth and industrial organization and the whole fabric and mechanism of social institutions—is to be regarded as a means to an end, and that end is the growth towards perfection of individual human beings.[1]

[1] R. H. Tawney, *Equality*, Allen & Unwin, 1964, p. 85.

Bibliography

Abegglen, J. G. *The Japanese Factory*. Free Press, 1958.
Allen, V. L. *Power in Trade Unions*, Longmans, Green & Co., 1954.
Allen, V. L. 'The Need for a Sociology of Labour'. *BJS*, 1959.
Allen, V. L. *Trade Union Leadership*, Longmans, Green & Co., 1957.
Allen, V. L. *Trade Unions and the Government*, Longmans, Green & Co., 1960.
Allen, V. L. *Militant Trade Unionism*, Merlin Press, 1966.
Anderson, N. *Work and Leisure*. Routledge and Kegan Paul, 1961.
Anderson, N. *Dimensions of Work: The Sociology of a Work Culture*. McKay, 1964.
Arendt, H. *The Human Condition*, Doubleday, 1959.
Argyris, C. *Personality and Organisation*, Harper & Row, 1957.
Argyris, C. *Understanding Organisational Behaviour*, Tavistock Publications, 1960.
Aron, R. *Eighteen Lectures on Industrial Society*, Weidenfeld & Nicolson, 1969.
Bain, G. S. *The Growth of White-Collar Unionism*. OUP, 1970.
Bakke, E. W. *The Unemployed Man*, Nisbet, 1933.
Bakke, E. W. *Citizens Without Work: A Study of the Effects of Unemployment Upon the Worker's Social Relations and Practices*, Yale University Press, 1940.
Baldamus, W. 'Type of Work and Motivation, in *BJS* 1951, II, pp. 44–51.
Baldamus, W. *Efficiency and Effort*, Tavistock Publications, 1961.
Banks, J. *Industrial Participation: Theory and Practice: a case study*, Liverpool University Press, 1963.
Banks, J. *Marxist Sociology in Action: a Sociological Critique of the Marxist Approach to Industrial Relations*, Faber & Faber, 1970.
Banks, O. *The Attitudes of Steelworkers to Technical Change*, Liverpool University Press, 1960.
Baran, P. A. and Sweezy, P. M. *Monopoly Capital*, Penguin, 1968.
Behrend, H. 'Absence and Labour Turnover in a Changing Economic Climate' in *Occupational Psychology*, 1953, XXVII, pp. 69–79.
Behrend, H. 'Voluntary Absence from Work' in *International Labour Rev.*, LXXIX, No. 2, February 1959.
Behrend, H. 'The Effort Bargain' in *Indus. and Lab. Rels. Review*, X, 1957.

Behrend, H. 'A Fair Day's Work' in *Scottish Jnl. of Pol. Economy*, VIII, June 1961.
Behrend, H. 'Price Images, Inflation and National Incomes Policy' in *Scottish Jnl. of Pol. Economy*, XIII, November 1966.
Bell, D. *The End of Ideology*, Free Press, 1960.
Bendix, R. 'Bureaucracy: The Problem and its Setting' in *ASR* XII, 1947, pp. 493–507.
Bendix, R. 'Industrialisation, Ideologies and Social Structures' in *ASR*, October, 1959, pp. 617–19.
Bendix, R. 'The Lower Classes and the Democratic Revolution' in *Industrial Relations*, I, No. 1, October 1961, pp. 91–116.
Bendix, R. *Work and Authority in Industry*, Harper & Row, 1963.
Bendix, R. *Nation-building and Citizenship*, John Wiley & Sons Inc., 1964.
Berle, A. and Gardiner, M. *The Modern Corporation and Private Property*, Macmillan, 1937.
Berle, A. A., Jn. *The Twentieth-Century Capitalist Revolution*, Harcourt Brace, 1954.
Berger, P. (ed.) *The Human Shape of Work*, Collier, 1964.
Bescoby, J. and Turner, H. A. 'Analysis of Post War Labour Disputes in the British Car-Manufacturing Firms', Manchester School, XXIX, 2. 1961.
Birnbaum, N. *The Crisis of Industrial Society*, Oxford University Press, 1969.
Blackburn, R. and Cockburn, A. (eds.) *The Incompatibles: Trade Union Militancy and the Consensus*, Penguin, 1967.
Blackburn, R. M. and Prandy, K. 'White Collar Unionisation: a Conceptual Framework' in *Brit. Jnl. Soc.*, XVI, 1965.
Blau, P. M. 'Co-operation and Competition in a Bureaucracy' in *AJS*, LIX, 1954.
Blau, P. M. *The Dynamics of Bureaucracy*, University of Chicago Press, 1955.
Blau, P. and Duncan, O. D. *The American Occupational Structure*, John Wiley & Sons Inc., 1967.
Blauner, R. 'Work Satisfaction and Industrial Trends in Modern Society' in Smelser, N. and Lipset, S. (eds.), *Labour and Trade Unionism*, John Wiley & Sons Inc., 1960.
Blauner, R. *Alienation and Freedom*, University of Chicago Press, 1964.
Blumberg, P. *Industrial Democracy: the Sociology of Participation*, Constable, 1968.
Blumer, H. 'Sociological Theory in Industrial Relations' in *ASR*, XII, 1947, No. 3, pp. 271–8.
Bottomore, T. B. *Elites and Society*, Watts, 1964.
Bottomore, T. B. *Classes in Modern Society*, George Allen & Unwin, 1965.
Boulding, K. *The Organisational Revolution*, Harper & Row, 1953.
Brady, R. A. *Business as a System of Power*, Columbia University Press, 1943.

Briggs, A. and Saville, J. (eds.)   *Essays in Labour History in Memory of G. D. H. Cole*, Macmillan, 1960.
Brown, D. V. and Myers, C. A.   *The Changing Industrial Relations Philosophy of American Management* Proc. of Ninth Annual Meeting of Industrial Relations Research Assn. Madison, Wisc., 1957.
Brown, R. K.   Research and Consultancy in Industrial Enterprises: a review of the contribution of the Tavistock Institute to the development of industrial sociology, *Sociology*, I, No. 1, 1967.
Brown, R. K.   'Participation, Conflict and Change in Industry' in *Soc. Review*, XIII, 1965, No. 3.
Brown, R. and Brannen, P.   'Social Relations and Social Perspectives among Ship-building workers' in *Sociology*, IV, Nos. 1 & 2, 1970.
Brown, W.   *Piecework Abandoned*, Heinemann, 1962.
Brown, W.   *Exploration in Management*, Heinemann, 1960.
Burnham, J.   *The Managerial Revolution: What is Happening in the World*, John Day, 1941.
Burns, T. (ed.)   *Industrial Man*, Penguin, 1969.
Burns, T. and Stalker, G.   *The Management of Innovation*, Pergamon Press, 1961.
Cameron, G. C.   'Post-War Strikes in the North-East Shipbuilding & Ship Repairing Industry' in *BJIR*, II, No. 1.
Cannon, I. C.   Ideology and Occupational Community: a study of Compositors in *Sociology*, I, No. 2, 1967.
Caplow, T.   *The Sociology of Work*, Manchester University Press, 1954.
Carey, A.   'The Hawthorne Studies. A Radical Criticism', *ASR*, XXXII, 3, June 1967.
Chamberlain, Neil, W.   *The Union Challenge to Management Control*, Harper & Row, 1948.
Chandler, Margaret K.   *Management Rights and Union Interests*, McGraw-Hill, 1964.
Child, A.   *British Management Thought: A Critical Analysis*, Allen and Unwin, 1969.
Child, J.   *Industrial Relations in the British Printing Industry*, George Allen & Unwin, 1967.
Child, J.   *The Business Enterprise in Modern Industrial Society*, Collier Macmillan, 1969.
Chinoy, E.   *Automobile Workers and the American Dream*, Doubleday, 1955.
Clack, G.   *Industrial Relations in a British Car Factory*, Cambridge University Press, 1967.
Clegg, H.   *A New Approach to Industrial Democracy*, Oxford University Press, 1960.
Clegg, H., Killick, A. and Adams, R.   *Trade Union Officers*, Basil Blackwell & Co., 1961.
Clegg, H.   *The System of Industrial Relations in Great Britain*, Blackwell, 1970.
Coates, K. and Topham, A. (eds.)   *Industrial Democracy in Great Britain*, McGibbon & Kee, 1968.

Coates, K. and Topham, A. (eds.) *Workers Control*, Panther, 1968.
Cole, G. D. H. *Workshop Organisation*, Oxford University Press, 1923.
Collins, D., Dalton, M. and Roy, D. 'Restriction of Output and Social Clearage in Industry' in *App. Anthropology* (now *Human Org.*) V, 1946.
Conquest, R. (ed.) *Industrial Workers in the U.S.S.R.*, Bodley Head, 1967.
Crozier, M. *The Bureaucratic Phenomenon*, Tavistock Publications, 1964.
Cunnison, S. *Wages and Work Allocation: a study of social relations in a Garment Workshop*, Tavistock Publications, 1965.
Curle, A. 'The Sociological Background to Incentives' in *Occupational Psychology*, XXIII, 1949, pp. 21–3.
Dahrendorf, R. *Class and Class Conflict in an Industrial Society*, Routledge & Kegan Paul, 1959.
Dalton, M. *Men Who Manage*, John Wiley & Sons Inc., 1957.
Dalton, M. 'The Industrial Ratebuster' in *Appl. Anthropology*, VII, 1948.
Dalton, M. 'Worker Response and Social Background' in *Jnl. of Pol. Econ.*, LV, August 1947.
Dennis, N., Henriques, F. and Slaughter, C. *Coal is Our Life*, Eyre & Spottiswoode, 1956.
Derber, M. Labour Management Relations at the Plant Level under Industry-wide Bargaining, *Institute of Labor and Industrial Relations*, University of Illinois, 1955.
Derber, M., Chalmers, W. E. and Edelman, M. T. Plant Union Management Relations: from Practice to Theory, *Institute of Labor and Industrial Relations*, University of Illinois, 1965.
Deutsch, S. 'The Sociology of the American Worker' in *Int. Jnl. of Comp. Sociology*, X, Nos. 1–2, March–June 1969.
Diebold, J. *Automation, the Advent of the Automatic Factory*, D. van Nostrand, 1952.
Djilas, M. *The New Class*, George Allen & Unwin, 1967.
Drucker, P. *Concept of the Corporation*, John Day, 1946.
Drucker, P. *The Practice of Managment*, Heinemann, 1955.
Dubin, R. 'Industrial Workers' Worlds: A Study of "Central Life Interests" of Industrial Workers' in *Social Problems*, III, 1956.
Dubin, R. Power and Union-Management Relations, *Administrative Science Quarterly*, II, 1957, pp. 60–81.
Dubin, R. *The World of Work: Industrial Society & Human Relations*, Prentice-Hall, 1958.
Dubin, R. *Working Union-Management Relations*, Prentice-Hall, 1958.
Dubin, R. 'Industrial Research and the Discipline of Sociology', *Industrial Relations Research Association. Proceedings of the 11th Annual Meeting*, 1959.
Dufty, N. F. (ed.) *The Sociology of the Blue-collar Worker*, E. J. Brill, 1969.
Dunlop, J. T. *The Theory of Wage Determination*, Macmillan, 1957.
Durkheim, E. *Professional Ethics & Civil Morals*, Routledge & Kegan Paul, 1957.

Durkheim, E.   *The Division of Labour in Society*, Free Press, 1964.
Durkheim, E.   *Socialism*, Collier, 1962.
Edelstein, J. D., Warner, M. and Cooke, W. G.   'The Pattern of Opposition in British and American Unions' in *Sociology*, IV, No. 2, 1970.
Eldridge, J. E. T.   *Industrial Disputes: Essays in the Sociology of Industrial Relations*, Routledge & Kegan Paul, 1968.
Faunce, W. A.   *Problems of an Industrial Society*, McGraw-Hill, 1968.
Feldman, A. J. and Moore, W. E.   'Industrialisation and Industrialism: Convergence and Differentiation' in Faunce, W. A. and Form, W. (eds.), *Comparative Perspectives on Industrial Society*, Little Brown & Co., 1969.
Fensham, P. J. and Hooper, D.   *The Dynamics of a Changing Technology: a case study in Textile manufacturing*, Tavistock Publications, 1964.
Finnegan, T. A.   'The Work Experience of Men in the Labour Force: an Occupational Study' in *Industrial and Labour Relations Review*, XVII, January 1964, No. 2, pp. 238–56.
Flanders, A.   *The Fawley Productivity Agreements*, Faber & Faber, 1964.
Flanders, A.   *Management and Unions. The Theory and Reform of Industrial Relations*, Faber & Faber, 1970.
Florence, P. S.   *The Logic of British & American Industrial Organisations*, Longmans, Green & Co., 1951.
Florence, P. S.   *Ownership, Control & Success of Large Companies*, Cambridge University Press, 1960.
Form, W. H.   'Occupational and Social Integration of Automobile Workers in Four Countries: a Comparative Study' in *Int. Jnl. of Comp. Sociology*, X, Nos. 1–2, 1969.
Form, W. H. and Miller, D. C.   *Industry, Labour and Community*, Harper & Row, 1960.
Fox, A.   'Managerial Ideology and Labour Relations' in *BJIR*, IV, No. 3.
Fox, A.   *Industrial Sociology and Industrial Relations*, Research Paper No. 3, Royal Commission on Trade Unions.
Fox, A.   *A Sociology of Work in Industry*, Collier-Macmillan, 1971.
Fraser, R. (ed.)   *Work* (2 vols.), Pelican, 1968, 1969.
Friedmann, G.   *Industrial Society*, Free Press, 1955.
Friedmann, G.   *Anatomy of Work*, Free Press, 1961.
Ginzberg, E. and Berman, H.   *The American Worker in the Twentieth Century*, Free Press, 1963.
Galbraith, J. K.   *The Affluent Society*, Hamish Hamilton, 1958.
Galbraith, J. K.   *American Capitalism: The Concept of Countervailing Power*, Houghton Mifflin, 1952.
Galbraith, J. K.   *The New Industrial State*, Houghton Mifflin, 1967.
Gellner, E.   *Thought and Change*, Weidenfeld & Nicolson, 1964.
Goldstein, J.   *The Government of British Trade Unions*, George Allen & Unwin, 1952.
Goldthorpe, J. H.   'Social Stratification in Industrial Society' in *Sociological Review Monograph*, No. 8, 'The Development of Industrial Societies', 1964.

Goldthorpe, J. H. 'Attitudes and Behaviour of Car Assembly Workers: a deviant case and a theoretical critique' in *Brit. Jnl. of Sociology*, XVII, September 1966.
Goldthorpe, J. H. et al. 'The Affluent Worker and the thesis of Embourgeoisement: some preliminary research findings' in *Sociology*, I, 1967.
Goldthorpe, J. H. et al. *The Affluent Worker*, Cambridge University Press, 1968–69, 3 vols.
Goldthorpe, J. H. and Lockwood, D. 'Affluence and the British Class Structure' in *Soc. Review*, 1963.
Goodrich, C. L. *The Frontiers of Control. A Study in British Workshop Relations*, Bell, 1920.
Gordon, R. A. *Business Leadership in the Large Corporation*, University of California Press, 1961.
Gouldner, A. W. *Wildcat Strike*, Routledge & Kegan Paul, 1955.
Gouldner, A. W. *Patterns of Industrial Bureaucracy*, Routledge & Kegan Paul, 1959.
Granick. D. *The Red Executive*, Macmillan, 1960.
Granick, D. *Management of the Industrial Firm in the U.S.S.R.*, Columbia University Press, 1954.
Granick, D. *The European Executive*, Weidenfeld & Nicolson, 1962.
Gross, E. *Work and Society*, Cromwell, 1958.
Gross, E. *Industry and Social Life*, W. C. Brown, 1965.
Guest, R. H. 'Work, Careers and Aspirations of Automobile Workers' in *ASR*, XIX, 1954, pp. 155–63.
Halker, A. 'The Use and Abuse of Pareto in Industrial Sociology' in *American Journal of Economics & Sociology*, XIV, 1955, pp. 321–34.
Hall, J. and Caradog, J. D. 'Social Grading of Occupations' in *BJS*, I, 1950.
Hamilton, R. *Affluence and the French Worker in the Fourth Republic* in Princeton University Press, 1967.
Harbison, F. and Myers, C. A. (eds.) *Management in the Industrial World: An International Analysis*, McGraw-Hill, 1959.
Hickson, D. J. 'Motives of Work of People who Restrict their Output' in *Occupational Psychology*, XXXV, July 1961.
Hobhouse, L. T. *Liberalism*, Oxford University Press, 1964.
Hobsbawm, E. J. *Labouring Men: studies in the history of labour*, Weidenfeld & Nicolson, 1964.
Hobsbawm, E. J. and Rudé, G. *Captain Swing*, Lawrence and Wishart, 1969.
Hollowell, P. *The Lorry Driver*, Routledge and Kegan Paul, 1968.
Homans, G. C. *The Human Group*, Routledge & Kegan Paul, 1950.
Hughes, E. *Men and Their Work*, Free Press, 1958.
Ingham, G. K. 'Organisational Size, Orientation to Work and Industrial Behaviour', *Sociology*, 1, 3, 1967.
Inkeles, A. 'Industrial Man: the Relation of Status to Experience, Perception and Value' in Landsberger, H. A. (ed.), *Comparative Perspectives on Formal Organisations*, Little, Brown & Co., 1970.

BIBLIOGRAPHY

Jacques, E. *The Changing Culture of a Factory*, Tavistock Publications, 1951).
Jaffee, A. J. and Carleton, R. O. *Occupational Mobility in the United States 1930–60*, Kings Crown Press, 1954.
Jeffreys, M. *Mobility in the Labour Market*, Routledge & Kegan Paul.
Kahn, H. *Repercussion of Redundancy*, George Allen & Unwin, 1965.
Kahn, R. L. and Morse, M. C. 'The Relationship of Productivity to Morale' in *Journal of Social Issues*, VII, 1951, pp. 8–17.
Kelly, J. *Is Scientific Management Possible?*, Faber & Faber, 1968.
Kerr, C., Dunlop, J. T., Harbin, F. H. and Myers, C. A. *Industrialism and Industrial Man*, Heinemann, 1960.
Kerr, C. *Labour and Management in Industrial Society*, Doubleday, 1964.
Knowles, K. G. J. C. *Strikes*, Basil Blackwell, 1952.
Kolko, G. *Wealth and Power in America*, Praeger, 1962.
Kornhauser, A. *Mental Health of the Industrial Worker*, John Wiley & Sons Inc., 1965.
Kornhauser, A., Dubin, R. and Ross, A. M. (eds.) *Industrial Conflict*, McGraw-Hill, 1954.
Kornhauser, W. *The Politics of Mass Society*, Routledge & Kegan Paul, 1960.
Lambert, R. *Workers, Factories and Social Change in India*, Princeton University Press, 1963.
Lane, T. and Roberts, K. *Strike at Pilkingtons*, Fontana, 1971.
Landsberger, H. *Hawthorne Revisited*, Cornell University Press, 1958.
Lerner, S. *Breakaway Unions and the Small Trade Union*, George Allen & Unwin, 1961.
Lerner, S. W. and Bescoby, J. 'Shop Steward Combine Committees in the British Engineering Industry' in *Brit. Jnl. of Ind. Relas.*, IV, No. 2, 1966.
Levenstein, A. *Why People Work*, Cromwell-Collier, 1962.
Lewis, W. A. *Theory of Economic Growth*, George Allen & Unwin, 1955.
Lipset, S. M. *et alia*. *Union Democracy*, Free Press, 1956.
Lipset, S. M. *Political Man*, Doubleday, 1960.
Lipset, S. M. and Bendix, R. *Social Mobility in Industrial Society*, University of California Press, 1959.
Lockwood, D. *The Blackcoated Worker*, George Allen & Unwin, 1958.
Lockwood, D. 'The "New Working Class"' in *European Jnl. of Sociology*, I, 1960, No. 2.
Lockwood, D. 'Sources of Variation in Working Class Images of Society' in *Soc. Review*, 1966.
Lupton, T. *On the Shop Floor*, Pergamon Press, 1963.
Lupton, T. and Cunnison, S. 'Workshop Behaviour' in Gluckman, M. and Devons, E. (eds.), *Closed Systems and Open Minds*, Oliver & Boyd, 1964.
Mann, F. C. and Hoffman, L. R. *Automation and the Worker*, Holt, Rinehart and Winston, 1960.

March, J. G. and Simon, H. A. *'Organisations'*, John Wiley & Sons Inc., 1958.
Marcson, S. (ed.). *Automation, Alienation and Anomie*, Harper & Row, 1970.
Marcus, P. 'Organisational Change: the Case of American Trade Unions' in Zollschan, G. K. and Hirsch, W. (eds.), *Explorations in Social Change*, Houghton Mifflin.
Marcuse, H. *One-dimensional Man*, Routledge & Kegan Paul, 1964.
Marriott, R. 'Size of Working Group and Output' in *Occupational Psychology*, XXII, 1949, pp. 47–57.
Marriot, R. 'Socio-Psychological Factors in Productivity' in *Occupational Psychology*, 1951, pp. 15–24.
Marriott, R. *Incentive Payment Systems*, Staples, Press Revised 1961.
Marsh, A. *Industrial Relations in Engineering*, Pergamon Press, 1965.
Martin, R. 'Union Democracy: an Explanatory Framework' in *Sociology*, II, No. 2, 1968.
Marx, K. *Economic and Philosophical Manuscripts of 1844*, Lawrence & Wishart, 1970.
Marx, K. and Engels, F. *The German Ideology*, Lawrence & Wishart, 1965.
Marx, K. *Capital*, London, 1889.
Mason, E. J. (ed.). *The Corporation in Modern Society*, Harvard University Press, 1959.
Matthewson, J. B. *Restriction of Output among Unorganised Workers*, Viking Press, 1931.
Mayo, E. *The Human Problems of an Industrial Civilization*, Harvard University Press, 1946.
Mayo, E. *The Social Problems of an Industrial Civilization*, Routledge & Kegan Paul, 1949.
McCarthy, W. E. J. *The Closed Shop in Britain*, Basil Blackwell, 1964.
McCarthy, W. E. J. *The Role of Shop Stewards in British Industrial Relations*, H.M.S.O., 1966.
McCarthy, W. E. J. and Parker, S. R. *Shop Stewards & Workshop Relations*, H.M.S.O., 1968.
McGeown, P. *Heat the Furance Seven Times*, Hutchinson, 1967.
McKenzie, R. T. and Silver, A. 'Conservatism, Industrialism and the Working Class Tory in England.' Transcription of the 5th World Congress of Sociology, III, Washington, 1962.
Mead, M. (ed.). *Cultural Patterns and Technical Change*, UNESCO, Paris, 1953.
Merton, R. K. 'Bureaucratic Structure and Personality' in *Social Forces*, XVIII, 1940, pp. 560–68.
Merton, R. K., Gray, A. P. Hockey, B. and Selvin, H. C. *Reader in Bureaucracy*, Free Press, 1952.
Merton, R. *Social Theory and Social Structure*, Free Press, 1957.
Michels, R. *Political Parties*, Collier, 1962.
Miliband, R. *The State in Capitalist Society*, Weidenfeld & Nicolson, 1968.

Miller, D. C. and Form, W. H. *Industrial Sociology. The Sociology of Work Organisations*, Harper & Row, 1964.
Miller, E. J. and Armstrong, D. 'The influence of advanced technology on the structure of management organisation' in Steiber, J. (ed.) *Employment problems of automation and advanced technology: an international perspective*, Macmillan, 1964.
Miller, E. J. and Rice, A. K. *Systems of Organisation*, Tavistock Publications, 1967.
Miiler, R. C. 'The Dockworker Sub-culture and some Problems in Cross-Cultural and Cross-Time Generalities' in *Comparative Studies in Society and History*, II, 3, 1969.
Mills, C. Wright. 'The Contribution of Sociology to Studies of Industrial Relations' *Ind. Rel. Research Ass.*, 1948.
Mills, C. Wright. *White Collar*, Oxford University Press, 1956.
Mills, C. Wright. *The Power Elite*, Oxford University Press, 1956.
Mills, C. Wright. *The New Men of Power*, Harcourt Brace, 1948.
Ministry of Labour. Final Report of the Committee of Inquiry under the R.H. Lord Devlin into certain matters concerning the Port Transport Industry. (Cmnd. 2734, H.M.S.O, 1965.
Moore, W. E. *Industrial Relations and the Social Order*, 2nd ed. Macmillan, 1951.
Moore, W. E. 'Current Issues in Industrial Sociology' in *ASR*, XII, 1947, No. 6, pp. 651–7.
Moore, W. E. *Industrialisation and Labour*, Cornell University Press, 1951.
Moore, W. E. *The Impact of Industry*, Prentice-Hall, 1965.
Mouzelis, N. *Organisation and Bureaucracy*, Routledge & Kegan Paul, 1967.
National Board for Prices and Incomes, Report 36, Productivity Agreements, H.M.S.O., 1966, Cmnd. 3311.
National Board for Prices and Incomes, Report 65, Payment by Results Systems, H.M.S.O., 1968, Cmnd. 3627.
National Board for Prices and Incomes, Report 123, Productivity Agreements, H.M.S.O., 1969, Cmd. 4136.
National Board for Prices and Incomes, *Hours of Work, Overtime and Shift Working*, H.M.S.O., 1971.
Nicholls, T. *Ownership, Control and Ideology: an Inquiry into Certain Aspects of Modern Business Ideology*, George Allen & Unwin, 1969.
Nordlinger, E. A. *The Working Class Tories*, McGibbon & Kee, 1967.
Nosow, S. and Form, W. H. (eds.). *Man, Work and Society*, Basic Books, 1962.
O.E.C.D. Steel Works and Technical Progress: A comparative Report of six national studies, O.E.C.D., Paris, 1959.
O.E.C.D. Adjustment of Workers to Technical Change at Plant Level, Int. Conference, O.E.C.D., 1966.
Osipov, G. V. (ed.). *Industry and Labour in the U.S.S.R.*, Tavistock Publications, 1966.
H

Palmer, G. L. *Labour Mobility in Six Cities*, Social Science Research Council, USA, 1954.
Parker, S. R. et al. *The Sociology of Industry*, George Allen & Unwin, 1967.
Parkin, F. 'Working-Class Conservatives: a theory of political deviance' in *BJS*, XVIII, No. 3, 1967.
Parsons, T. and Smelser, N. J. *Economy and Society*, Routledge & Kegan Paul.
Paterson, T. *Glasgow Ltd.—a case study in industrial war and peace*, Cambridge, 1960.
Perlman, S. *A Theory of the Labour Movement*, University Press, Macmillan, 1928.
Phelps-Brown, E. H. *The Economics of Labour*, Yale University Press, 1962.
Phelps-Brown, E. H. *The Growth of British Industrial Relations*, Macmillan, 1959.
Pelling, H. *History of Trade Unionism*, Pelican, 1964.
Pollard, S. 'Factory discipline in the Industrial Revolution' in *EHR*, 2nd Series, XVI, 1963.
Pollard, S. *The Genesis of Modern Management*, Penguin, 1968.
Pope, L. *Millhands and Preachers*, Yale University Press, 1965.
Prandy, K. (1965) *Professional Employees: A Study of Scientists and Engineers*, Faber, 1965.
Rice, A. K. *Productivity and Social Organisation: the Ahmedabad Experiment*, Tavistock Publications, 1958.
Riesman, D. *Abundance for what?* Doubleday, 1964.
Roberts, G. *Demarcation Rules in Shipbuilding and Shiprepairing*, Cambridge University Press, 1967.
Robertson, D. J. *Factory Wage Structures*, Cambridge University Press, 1960.
Roethlisberger, F. J. and Dickson, W. J. *Management and the Worker*, Harvard University Press, 1939.
Rose, A. M. 'The Potential contribution of Sociological Theory and Research to Economics' in *American Journal of Econs. & Sociology*, XII, 1952, pp. 23–33.
Ross, A. M. and Hoffman, P. T. *Changing Patterns of Industrial Conflict*, John Wiley and Sons Inc., 1960.
Routh, G. *Occupation & Pay in Great Britain*, 1906–60. Cambridge University Press, 1965.
Roy, D. F. 'Do Wage Incentives Reduce Costs?' in *Industrial Labour Relations Rev.*, V, 1952, pp. 195–201.
Roy, D. 'Quota Restriction & Goldbricking in a Machine Shop' in *AJS*, LVII, March 1952.
Roy, D. 'Work Satisfaction and Social Reward in Quota Achievement: an Analysis of Piecework Incentive' in *ASR*, XVIII, October 1953.
Roy, D. 'Efficiency and the "Fix": Informal inter-group Relations in Piecework Machine Shops' in *AJS*, LX, 1954.

Royal Commission on Trade Unions and Employers Associations, 1965–68, HMSO, 1968, Cmd. 3623.
Runciman, W. G. *Relative Deprivation and Social Justice*, Routledge & Kegan Paul, 1966.
Sayles, L. *Behaviour of Industrial Work Groups*, John Wiley & Sons Inc., 1958.
Schneider, E. V. *Industrial Sociology*, McGraw-Hill, New York, 1957.
Schumpeter, J. A. *Capitalism, Socialism and Democracy*, Harper & Row, 1947.
Scott, W. H. et al. *Technical Change and Industrial Relations*, Liverpool University Press, 1956.
Scott, W. Automation and the Non-Manual Worker—interim report on a comparative project in 5 European countries, OECD Paris, 1962.
Scott, W. H., Mumford, Enid, McGivering, I. C. and Kirkby, J. M. *Coal & Conflict, a study of Industrial Relations at Collieries*, Liverpool University Press, 1963.
Sheppard, H. L. 'The Treatment of Unionism in "Management Sociology" ', *ASR*, XIV, 1949.
Sheppard, H. L. 'Approaches to Conflict in American Industrial Sociology' in *BJS*, V, 1954, pp. 324–41.
Shimmin, Sylvia, 'Extra-mural factors influencing behaviour at work' in *Occupational Psy.*, XXXVI, 1962.
Shonfield, A. *Modern Capitalism*, Oxford University Press, 1965.
Shostak, A. B. and Gomberg, W. (eds.). *Blue-collar World*, Prentice-Hall, 1964.
Silverman, D. *Theory of Organisations: A Sociological Framework*, Heinemann, 1970.
Simey, T. (ed.). *The Dockworker*, Liverpool University Press, 1954.
Smelser, N. J. *Social Change in the Industrial Revolution*, Routledge & Kegan Paul, 1959.
Smelser, N. J. *The Sociology of Economic Life*, Prentice-Hall, 1963.
Smelser, N. J. *Readings on Economic Sociology*, Prentice-Hall, 1965.
Smelser, N. J. and Lipset, S. M. (eds.). *Social Structure and Mobility in Economic Development*, Aldine Press, 1966.
Sorel, G. *Reflections on Violence*, Collier Books, 1961.
Sorenson, R. C. 'The Concept of Conflict in Industrial Sociology' in *Soc. Forces*, XXIX, 1951, No. 3, pp. 263–7.
Steiber, J. (ed.). *Employment Problems of Automation and Advanced Technology—an International Perspective*, Macmillan, 1966.
Stone, R. C. 'Conflicting Approaches to the Study of Worker–Manager Relations' *Soc. Forces*, VI, 1952, 31, pp. 117–24.
Sturmthal, A. (ed.). *White Collar Trade Unions*, University of Illinois Press.
Sturmthal, A. 'Industrial Democracy in the Affluent Society', *Industrial Relations Research Association. Proceedings of the 7th Annual Meeting*, 1964.
Sykes, A. J. M. 'Unity and Restrictive Practices in the British Printing Industry' in *Soc. Rev.*, VIII, 1960.

Sykes, A. J. M. 'Trade Union Workshop Organisation in the Printing Industry' in *Human Relations*, XIII, 1960.
Sykes, A. J. M. 'A Study in Changing the Attitudes and Stereotypes of Industrial Workers' in *Human Relations*, XVII, 1964.
Sykes, A. J. M. 'Some Differences in the Attitudes of Clerical and Manual Workers' in *Sociological Review*, XIII, 1965.
Sykes, A. J. M. 'An Industrial Rite de Passage' in *Man*, May 1965.
Sykes, A. J. M. 'Joking Relationships in an Industrial Setting' in *American Anthropologist*, LXVIII, 1966.
Sykes, A. J. M. 'The Cohesion of a Trade Union Workshop Organisation' in *Sociology*, I, No. 2, 1967.
Sykes, A. J. M. 'Navvies: Their Work Attitudes' in *Sociology*, III, No. 1, January 1969.
Sykes, A. J. M. 'Navvies: Their Social Relations' in *Sociology*, III, No. 2, May 1969.
Tannebaum, A. S. 'Control Structure & Union Functions' in *AJS*, LXI, 1956, pp. 536–45.
Taussig, F. W. and Joslyn, C. S. *American Business Leaders: A Study in Social Origins and Social Stratification*, Macmillan, 1932.
Tawney, R. H. *The Acquisitive Society*, Fontana, 1961.
Tawney, R. H. *Equality*, George Allen & Unwin, 1964.
Taylor, F. W. *Principles of Scientific Management*, New York, 1911.
Taylor, F. W. *Scientific Management*, Harper, 1947.
Taylor, F. W. *Piece Rate System*, Routledge & Kegan Paul, 1919.
Thompson, E. P. *The Making of the English Working Class*, Victor Gollancz, 1963.
Thompson, E. P. 'Time Work Discipline and Industrial Capitalism' in *Past and Present*, December 1967.
Touraine, A. et al. *Workers Attitudes to Technical Change*, O.E.C.D. Paris, 1965.
Trist, E. C. and Bamforth, K. W. 'Some Social & Psychological Consequences of the Longwall Method of Coal-jetting' in *Human Relations*, IV, 1951.
Trist, E. L. et al. *Organisational Choice: Capabilities of Groups at the Coal Face under Changing Technologies*, Tavistock Publications, 1963.
Tunstall, J. *The Fishermen*, McGibbon and Kee, 1962.
Turner, H. A. *Trade Union Growth Structure and Policy. A comparative Study of the Cotton Unions*, George Allen & Unwin, 1962.
Turner, H. A. *The Trend of Strikes*, Leeds University Press, 1963.
Turner, H. et al. *Labour Relations in the Motor Industry*, George Allen & Unwin, 1967.
Turner, H. A. *Is Britain really strike-prone?*, Cambridge University Press, 1969.
Turner, H. A. and Bescoby, J. 'Strikes, Redundancy and the Demand Cycle in the Motor Car Industry', *Oxford Bulletin*, XXIII, No. 2, 1961.
Veblen, T. *The Theory of Business Enterprise*, Scribner, 1904.
Veblen, T. *Engineers and the Price System*, B. W. Huebsch, 1921.
Veblen, T. *The Theory of the Leisure Class*, Macmillan, 1917.

Veblen, T. *The Instinct of Workmanship*, Norton, 1941.
Walker, Charles R. *Steeltown: An Industrial Case History of the Conflict between Progress & Security*, Harper & Row, 1950.
Walker, C. R. *Towards the Automatic Factory*, Yale University Press, 1957.
Walker, C. R. and Guest, R. H. *The Man on the Assembly Line*, Harvard University Press, 1952.
Walker, C. R. and Walker, A. G. (eds) *Modern Technology & Civilization: An Introduction to Human Problems in the Machine Age*, New York, McGraw-Hill Book Co. Inc., 1962.
Warner, W. L. and Low, J. W. *The Social System of a Modern Factory*, Yale University Press, 1946.
Webb, S. and B. *History of Trade Unionism*, London, 1894.
Webb, S. and B. *Industrial Democracy*, London, 1811.
Weber, Max *The Protestant Ethic and the Spirit of Capitalism*, George Allen & Unwin, 1930.
Weber, Max *Economy and Society* (3 vols.), Bedminster Press, 1968.
Wedderburn, D. *White Collar Redundancy*, Cambridge University Press, 1964.
Wedderburn, D. *Enterprise Planning for Change*, O.E.C.D., 1968.
Wedderburn, K. *The Worker and the Law*, Penguin, 1971.
Whyte, W. F. *et al.* *Money and Motivation*, Harper & Row, 1955.
Whyte, W. H. *The Organisation Man*, Simpson & Slimster, 1956.
Wilkie, R. 'The Ends of Industrial Sociology' in *Soc. Rev.* July 1961, pp. 215-24.
Willener, A. 'Payment Systems in the French Steel & Iron Mining Industry: An Exploration in Managerial Resistance to Change' in Zollschan, G. K. and Hirsch, W., *Explorations in Social Change*, Houghton & Miffin, 1964.
Woodward, J. *Industrial Organisation: Theory and Practice*, Oxford University Press, 1965.
Woodward, J. *Industrial Organisation: Behaviour and Control*, Oxford University Press, 1970.
Wootton, B. *The Social Foundations of Wage Policy*, George Allen & Unwin, 1955.
Zeitlin, M. *Revolutionary Politics and the Cuban Working Class*, Princeton University Press, 1967.
Zweig, F. *The British Worker*, Penguin, 1952.
Zweig, F. *Productivity and the Trade Unions*, Basil Blackwell, 1951.
Zweig, F. *The Worker in an Affluent Society*, Heinemann, 1961.

Index

Absenteeism, 47-8, 56, 59, 107
Acquisition, spirit of, 173
Action, 'actual' and 'types of', 43-4. *See also* Social action
Affluent workers, 54, 64-5, 67-8
Alienation, concept of, 9, 80, 135; Blauner's study of, 183-95; four dimensions of, 183-4, 190-2; as image of man's essential nature, 146; inverted-U thesis of, 185, 189, 192; Marx's analysis of work and, 139-58; Sartre's interpretation, 158-63; solution to 143; wages and private property as consequences of, 141; Weber's treatment of, 165-74; work satisfaction within, 195
Allen, V. L., 178, 182
Aluminum Corp. of America, 52
American Indians, 129
American Occupational Structure, The, 105
Anarchy, 90, 116
Andreski, S., 48
Anomia, 134
Anomie: community breakdown as, 100-11, 207; confused with alienation, 80; cure for, 85-91, 117, 192; definition of, 134; of deprivation, 134; Durkheimian tradition of, 9, 73-91, 134; in industrial relations, 112; as interpretation of economic life, 73-91; Merton's study of, 119-35; in 19th-century Europe, 81; of success, 81, 134

Appropriation: of labour, 165-7; of material means of production, 167; Weber's concept, 165-8
Arendt, H., 167n.
Argentina, study of car workers, 18-20
Ascetism, 172-3
Automobile workers. *See* Car workers
Automobile Workers and the American Dream, 127
Avineri, S., 156n., 170

Bad faith, concept of, 158, 160-1, 163
Bakke, E. Wight, 128, 129n.
Baldamus, W., wage-effort paradigm, 58-61
Bank Wiring Observation Room study, 29-30, 62-3, 108
Bargaining, growth of, 49
Barker, C., 133n.
Bay Chemical Co., 189
Becker, Howard S., 131
Behrend, H., 58
Bell, Daniel, 52, 109-10, 154, 205
Bell, Gerald D., 42n.
Bendix, Reinhard, 49n., 99
Berger, P. L., 39n., 42n., 160-4
Blau, P. M., 125; and Duncan, 105, 129
Blauner, Robert, *Alienation and Freedom*, 3, 19, 183-95
Bonus scheme, 58
Bridlington principles, 180

Bristol Siddeley Engines Ltd., 122–3
British Steel Corp. (BSC), 38–9
Bureaucracy, Marx on nature of, 156
Bureaucratic control (in trade unions), 175–82
Bureaucratization, modern phenomenon of, 170–2
Bureaucrats, ritualistic behaviour of, 124–7
Burnham, J., *Managerial Revolution*, 170
Burns, T., 62, 157

Cab-drivers, study of, 42
Café waiter, role of, 158–60
Caliente, 100
Cannon, I. C., 68–9
Cantril, H. W., 15n.
Capital, see under Marx
Capitalism: capitalist accumulation (Marx), 152, 157; growth of, 48, 98, 147, 154, 169, 174; social conditions under, 149–51
Captain Swing, 45
Car workers, comparative studies, 18–20, 66, 187–8, 191–3
Casual employment, effects of, 17–18, 48
Centers, Richard, 15n.
Change: attitudes to, 64, 89, 91–111; community breakdown through, 100–3; in industrial societies, 24; as source of disorder, 112–13
Chemical industry, 187, 189–90
Chicago, 30, 61, 106; school of sociologists, 41–4
Child, J., 170n.
Chinoy, E., 19–20, 127
Citizenship, concept of, 97–9
City life, anomic character of, 105–10, 149
Civil rights, 98–9
Class, development of, 148, 171
Clinard, M. B., 134n.
Cohen, Percy S., 188n.

Coleman, D. C., 47n.
Collective bargaining, 11–12, 116–17, 192
Collins, O., 54n.
Commission on Industrial Relations, 116
Communism: Djilas's view of, 157; Marx's comments on, 143–4, 146, 158; Weber's doubts, 170
Communist Manifesto (Marx), 150, 153
Community and association as forms of social order, 89; cases of breakdown of, 100–11
Complaints, scheme for interpretation of, 26–8
Compositors, study of, 69
Comte, Auguste, 92
Conflict: Durkheim's views on, 78–90, 109; German avoidance of, 200–1; management-union, 114
Conformity (Merton's category), 120, 133
Conquest, Robert, *The New Class*, 158
Conservative voters, 67–8
Consultants, management, 32, 36, 39, 107
Contract law, 84, 90
Cooper, D., 202
Co-operatives, workers', 154
Cottrell, study of community breakdown, 100–1
Craftsmanship, ideal typical, 194–5
Craftsmen, medieval, 87–8, 148, 167
Crowley, Thomas, 47
Crozier, M., 125–7
Cultural goals, 119–22
Cunnison, Sheila, 30–2

Dahrendorf, R., 96n., 157n.; concept of integration, 199–201; *Essays in the Theory* of Society, 199–200; *Society and Democracy in Germany*, 200

Dalton, M., 54, 123–4; *Men Who Manage*, 62, 123
Dance Musicians, 131
Davis, Fred, 42
Death by Dieselisation, 100
Deflationary fragmentation, 114
Dennis, N., 33, 66
Devlin, Lord, 49n.
Devlin Report, 48–9
Dickson, W. J., 26–9, 28–9, 108, 110
Disequilibrium, temporary, 27
Disorder: consequences of, 114–16; four sources of, 112–13
Division of labour: abnormal forms of, 76–83; anomic, 76–7, 82; degrading effect of, 77; forced, 78–80, 83, 135; international, 148; in towns, 147–8; Marx's concern with, 146–52; spontaneous, 78; views of political economists, 142; Weber's discussion on, 165
Djilas, M., 157–8
Dockworker sub-culture, 16–18
Donovan, Lord, 112
Donovan Report, 6n., 63–4, 112, 117
Dubin, Robert, 179, 191
Duncan, O. D., 105, 129
Dunlop, J. T., 2n.
Dunning, W., 50
Durham mining villages, 101
Durkheim, Emile, 3, 22, 92–7, 103–5, 109–12, 117, 120–1, 191–2; on abnormal forms of division of labour, 76–83; anomie theme, 9, 73–91, 133–5; framework of rules for industry, 89; mechanical solidarity, use of, 73–6, 120; notion of progress, 91; on social reconstruction and morality, 78–80, 83–91; works cited: *Division of Labour in Society*, 73, 80, 82–3, 103; *Professional Ethics and Civic Morals*, 87–8; *Socialism*, 83, 85, 89, 90; *Suicide*, 80–3, 91

Eastern Europe, de-casualization in, 17
East India Co., 156
Economic life: Durkheim on, 81, 83–91, 104; Goldthorpe on, 118
Economic men, 57–8
Ecstasy, Berger's definition, 164
Education in industrial societies, 23–4, 104–5, 177
Effort bargain, 49–63
Eldridge, J. E. T., 1, 3, 51n., 171n.
Electrical Engineering case study, 30
Elias, Norbert, 13
Employers' associations, 115
Employment and Manpower, Senate Sub-Committee, 129
Engels, F., 147–50ns.; *Conditions of English Working Class in 1844*, 151–2
Engineering industries, 50, 52–3, 59–60
Environment, 29, 31
Equilibrium, workers', 27–8
Europe, 19th-century, 81, 85, 106
Expropriation of workers and managers, 168

Factory life and systems, 23, 30–2, 77, 171, 183
Farms, machine-breaking on, 45
Faunce, W. A., 23n.
Feldman, Arnold S., 23–4
First World War, 114
Flanders, A., 112–19, 189
Ford, Henry, 51
Form, William H., 18–19, 25n.
Fortune magazine, 188
Fox, A., and Flanders, A., 'Reform of Collective Bargaining . . .', 112–19
France: provincial administration, 126; socialist workers, 144
Fraser, Ronald, 40n.
Freedom; alienation and, 183–95; personal, 162–4
French Economists, 93

INDEX

Galbraith, J. K., 175
Garment manufacture study, 30
General Motors Corp., 154
Germany: comparison of job satisfaction, 14–15; dislike of industrial disputes, 200–1; growth of clerical occupations, 171
Glacier Metal project, 32
Glasgow, 49
Glass workers, 132, 180
Gluckman, M., 31n., 32n.
Goffman, E., 161–3
Gold, Raymond L., 41–2
'Goldbricking', 56–7
Goldstein, J., 182
Goldthorpe, J. H., 19, 23, 32, 54n., 64–8, 117–18, 193n.
Gotha Programme, 153
Gouldner, A. W., 56, 89, 206–7
'Gravy jobs', 57
Guild system, 87–9, 148
Gypsum mining strike, 56

Harmony, concept of, 96; social, 200
Hawthorne investigations, 29–30, 61, 108, 110
Hegel, Georg W. F., 143, 146
Hobbes, Thomas, 91
Hobhouse, L. T., 92–7; on determination of wages, 94–5; forms of management, 96; organic nature of society, 92; protective legislation for manual labourer, 93
Hobsbawm, E. J., 45–6
Homans, G., *The Human Group*, 29–30
Horton, J., 133–4
Hughes, E. C., 43

Incomes policy, 117–18
India, 17–19, 156
Individual, social integration of. *See* Integration
Individual adaptation, modes of, 120–33
Individual approach, 40, 50–5

Individualism, excessive, 81, 86
Industrialism and Industrial Man (Kerr), 21
Industrial relations, 11–19; disorder in, 113–15; role of law of in, 116–17; solutions suggested, 116–18
Industrial societies: American, Mills's view of, 201; bureaucratization of, 170–2; integration of individual, 85, 199; position of managers in, 168–70; problem of anomie, 87, 91, 106; role of intellectuals in, 21; role of political activity, 24; similarities and differences of, 20–4; 'totalitarian' (Marcuse), 202
Industrial sociology, theoretical problem of, 193–4
Industries, Blauner's comparative chart, 186–7
Inflation, 114–15
Inkeles, Alan, 13–14, 16
Innovation, 120–7, 134
Institutional norms, 119–20
Integration, 119, 199–202; Durkheim's solution, 75; loss of, 113–14; Marcuse's theory, 202–7; Mayo's views, 107–9; 'social', 90, 109–10, 116–17; 'system', 85, 104, 111, 115
Isolation, workers', 186–7, 191, 193
Italy, 15, 18–19

Jacques, E., 33
Janitor (job study), 41–2
Jays Engineering study, 53
Jessop process, 123
Job, loss of, 128–9
Job analysis, 41–3
Job evaluation schemes, 115
Job satisfaction, 14–15, 18–20, 128, 201; engineered by managers, 195; of supervisors, 110
Journalism, 40–1

Kerr, Clark, 20–4, 132, 191, 202
Killingworth, Charles, 129–30

Kolko, G., 170n.
Komarovsky, M., 129n.
Kornhauser, A., 179n., 191
Kornhauser, W., 130

Labour: alienation of, 140, 146; appropriation of (Weber), 165–7, 170; division of, *q.v.;* estranged, 139–43; guild and non-guild, 148
Latin America, 48
Law, role of, 116–17
Law Book of the Crowley Iron Works, 47
Lazarsfeld, P., 129n.
Lenin, Nikolai, 49
Le Play, F., 2, 3
Lewis, John L., 178–9
Liberal socialism, 92–100
Lipset, S. M., 15n., 178–9, 205
Liverpool, 49
Lockwood, D., 64–8
'Logic of efficiency' and 'logic of sentiments', 61–2
London Typographical Soc., 181
Low, J. O., 101–4
Luckmann, T., 160n., 161
Lupton, T., 30–2, 53, 55
Luton, study of affluent workers, 54, 65–6, 68

Machine-breaking riots, 45–6
MacIntyre, A., critique of Marcuse, 205–7
Malthus, Thomas, 93, 155
Management: behaviour, 34–7, 107–8; causes of conflict with workers, 58; modes of, 96; separation from ownership, 168–70
Management and the Worker, 26, 61
Management Innovation, The, 62
'Managerial sociology', 109
Managers, 35–7, 62, 170
Manipulation, 202–3
Marcson, S., 130n.
Marcuse, Herbert, 202–7

Market economy, modern, 169
Market societies, 23, 199
Marriott, R., 19n.
Marshall, Alfred, 97
Marshall, T. H., 97–9
Martin, Roderick, 180
Marx, Karl, 2, 3, 22, 103, 135, 160, 164–5, 168–9, 173, 193–4; on 'abolition of labour', 204; alienation theme, 9, 139–58; on bureaucracy, 156; on communism, 143–4; on division of labour, 142–50; on forms of domination, 149; on piece-rate system, 50; on private property, 142–50; on wages, 141–2; works cited: *Capital*, 151–2, 154–5, 169; *Communist Manifesto*, 150, 153; *Critique of the Gotha Programme*, 153; *Economic and Philosophical MSS*, 139, 142–3, 146, 149, 151; *German Ideology*, 146–7; *Grundrisse*, 155; *Theorien*, 155, *Wage Labour and Capital*, 152
Mayo, Elton, 106–11
Mead, Margaret, 47
Meaninglessness, 80–2, 86, 186–7, 193
Mechanization, increase in, 152, 156, 171, 173
Men Who Manage, 62, 123
Merton, R. K., 3, 64–5, 106; anomie paradigm of, 119–27, 131–5; 'Bureaucratic Structure and Personality', 125; *Social Structure and Anomie*, 119; success theme, 120
Mica splitting investigation, 108
Michels, R., 175–7, 179n., 182
Mill, John Stuart, 94, 142
Miller, D. C., 25n.
Miller, E. J., 34–8ns.
Miller, R. C., 16–20
Mills, C. W., 3, 179, 193–5, 201–2
Mining Industry, 33–4, 66, 101, 129; 'seam society' of, 33
Money, role of (Marx), 142, 149

Moore, Wilbert E., 23–4
Myers, C. A., 21n.
Myth of General Strike, 132

National Board for Prices and Incomes (NBPI), 52–4, 59–60, 116
National Union of Mineworkers, 33
Natural history approach, 41
Navvies, study of, 54
Negroes, 129
New Class, The, 157n., 158
New Mexico, 47
Nicolaus, M., 146, 155–7
Normlessness, 193
Norway, analysis of job satisfaction, 15
Novel writing, 40

Occupational associations, 76, 88–9, 96, 111
Oil industry, USA, 189
Oligarchy, iron law of, 175, 177
One Dimensional Man, 202, 207
On the Shop Floor, 53, 58
Opportunity, inequalities of, 105, 117–18
Organization, systems of, 34–9; bureaucratic, 175–8
Orwell, George, 129
Over-conformity, 124–6
Ownership, divorced from control, 169–70

'Pacification of existence' (Marcuse), 204–6
Park, Robert E., 41
Payment by Results systems (PBR), 52–4, 59, 61
Phelps-Brown, E. H., 56
Piece-rate system, 50–7, 154–5
Pilkington Glass Works, 132–3, 180
Political economy, classical, 145–6

Pollard, S., 50
Popper, K., 23
Power, concept of, 165
Powerlessness, 186–7, 190–3
Printers, 185–6, 188–9
Private property, Marx's concern with, 141–50
Production, control of means of, 167–70
Productivity agreements, 60–1, 115, 189
Profit, 169
Property: Durkheim's views on, 84–90
Puerto Ricans, 129
Puritans, work ethic, 172–4
Purposelessness, concept of, 82

'Quota Restriction and Goldbricking in a Machine Shop', 56
Quota restrictions, 56–8

'Rabble hypothesis', 107
Ragged Trousered Philanthropists, 40
Rate-busters, 54
Rebellion, 120–1, 131–4
Reference group analysis, 64–9
Reification, 160–1
Relay Assembly Test Room studies, 108
Restrictive practices, 45–50, 55–64
Retreatism, 120–2, 134; of industrial workers, 127–8; of socially disconnected, 128–30
Ricardo, David, 156
Rice, A. K., 34–6, 37n.
Ritualism, 120–1, 134; of bureaucrats, 124–7
Roethlisberger, F. J., 26–9, 61–2, 63n., 108, 110
Role distance, concept of, 161–3
Role models, 159
Roper surveys, 19, 188
Ross, A. M., 179n., 191n.
Rossendale, 46
Rossi, P., 14, 16n.

Roth, G., 166n., 170n.
Roy, Donald, 53, 54n., 56–8, 124
Rudé, George, 45
Runciman, W. G., 64, 67–8

Saint-Simon, Claude Henri de, 84
Sartre, Jean-Paul, 158–63
Say, Jean–Baptiste, 142
Schneider, E. V., 25
Schumpeter, J. A., 175
Seaman's Ca'canny (1896), 55
Secondary groups, 86–7, 97
Second World War, 56, 114, 154
Segmentary relationships, 73–6
Self-estrangement, 161, 163–4, 173, 184, 186–7
Sentient system, definition of, 34
Shaw, G. B., 154
Ship-building, 129
Shoe industry (USA), 101
Shonfield, A., 63
Shop stewards, 53
Siegel, Abraham, 132, 191
Sinfield, A., 129n.
Smelser, N. J., 100n.
Smith, Adam, 93, 142
Smith, Ron, 38–9
Social action, 28, 32, 40–64; four types of (Weber), 43–5; illustration of, 45
Socialism, Weber's discussion of, 170
Social reconstruction, Durkheim's views, 83–91
Social Structure and Anomie, 119
Social System of the Modern Factory, 101–2
Social systems, prerequisites for, 25
Societies, different types of, 74, 107, 199, 203
Society, Marx's definition, 144
Sociology: concepts and aims, 207–8; methods of analysis, 40, 43–5
Socio-technical system, 33

Solidarity: mechanical, 73–6, 106; organic, 73–9, 85, 103, 106; role of, 80
Sorel, G., 132
Species character, 141
Spencer, Herbert, 22, 74
Spengler, Oswald, 106
Spontaneity, Durkheim's argument, 78
Srole, L., 134
Stalinism, 157
Stalker, G., 62
Steel industry, 34–9, 58, 129
Strategy-mix, 37–9
Stratification system, 23–4
Strikes, 56, 59, 101, 114, 132, 191, 201; union leaders and, 176, 182
Success goal, 120–1, 133
Suicide, anomic, egoistic and fatalistic, Durkheim on, 81–3
Surplus value theory, 153, 155
Sutherland, E. M., 122
Sweden, 15
Sykes, A. J. M., 54
System integration, 90, 104, 115
Systems analysis, 25–39; in Hawthorne studies, 29; Homan's three concepts, 29; in Tavistock research, 33–7

Tavistock Institute, 32–7
Tawney, R. H., 94, 208
Taylor, Frederick, 51–3
Technology: Blauner's four types, 184, 186–7, 193; demands of, in industry, 21; important effects of, 184–5
Textile workers, 148, 186, 192
Thompson, E. P., 47n., 49, 55
Time: attitudes to, 47–9; bargaining for, 49–55
Times, The, 45
Tönnies, F., 2, 106
Totalitarianism, 202–3
Trade Boards, 95
Trade unions, 18, 20, 32–3, 98–9, 110, 114–15, 191; aim of, 178;

INDEX 229

Trade unions—(contd.)
American, 154, 176, 178–9; British, 176–8, 180–2; and bureaucratic control, 175–82; closed and open, 181; membership fluctuations, 178–80
Traditional societies: attitude to time, 47–9; sympathy for machine-breakers, 45–6
Traditional workers, two types, 66–8
Tresswell, R., 40
Trist, E., 33
Trotsky, Leon, 157
Turner, H. A., 181–2

Under-employment, voluntary, 47
Unemployed Man, The, 128
Unemployment, 128–30
United Kingdom: conditions in 1844, 151–2; contemporary industrial relations in, 54–5, 112–19 (*see also* Trade unions); disappearance of sub-culture, 17; dissatisfaction with work tasks, 19; mining industry, *q.v.*; productivity agreements, *q.v.*;
United Mine Workers, 178–9
USA: Blauner's study of industrial life in, 183–95; community breakdown in, 100–5; effects of city life, 105–6; effects of industrialisation, 17–18; jazzmen reject society in, 131; oil industry, 189; piece-work systems, 51, 53n., 54; pursuit of wealth (Weber), 174; rate-busters, 54; Spanish Americans of, 47–8; study of car workers, 18–19; study of success in (Merton), 120; survey of job satisfaction, 14–16; system of control, 201–2; trade union leadership, 178–9; unemployment problem, 129–30
USSR: campaign for methodical work, 49; comparison of job satisfaction, 15; sociology of industry in, 206

'Unlimited desires', 80–1, 85
Unpunctuality, 48–9

Veblen, T., 167n.

Wage-Effort paradigm (Baldamus), 58–61
Wages, 113, 115–16; 'decaying' systems, 59–60; determination of, 59, 94–6, 116; Marx's views on, 141–2, 152–5
Wales, depopulated villages, 101
Walker, C. R., 19n.
Walker, J., 19n.
Warner, W. Lloyd, 101–4
Wealth: abrupt growth in, 80–1; hereditary, 79, 89–90; production and distribution of, 115; social inequalities of, 93, 111
Webb, Sidney and Beatrice, 177–8
Weber, Max, 2, 3, 40, 43–5, 194; on alienation, 165–74; on bureaucratization, 170–2; classification of division of labour, 165; concept of power, 165; on development of capitalism, 174; works cited: *The Protestant Ethic and the Spirit of Capitalism*, 47, 172, 174; *Theory of Social and Economic Organization*, 40, 44, 172; *Economy and Society*, 166, 170
Welfare state, 199, 203, 205
'Welting' or 'spelling', 48–9
Western Electric investigation, 61, 109
White collar, 194–5
Whitehead, A. N., 9
Whitley Council, 95
Wildcat Strike, 56
Wittich, C., 166n., 170n.
Woodward, J., 182n.
Work (R. Fraser), 40
Work and its Discontents, 52
Work ethic, puritan and utilitarian, 173–4

Work satisfaction, *see* Job satisfaction

Workers: activity of, 141; affluent, *see* Affluent workers; appropriation by, 167; as capital, 143; expropriation of, 167–8; women's attitudes, 54–5

Working class: civil rights movement, 98–9; Conservation of, 67–8; embourgeoisement of, 64–9; Marx and Engels on conditions in 1844, 151–2; oligarchical tendencies, 175–7

Wye Garment Co., 55

Yankee City shoe factory, 101–4

Zawadski, B., 129n.
Zola, E. *Germinal*, 40